T0271151

# Numeral Systems with Irrational Bases for Mission-Critical Applications

K&E Series on Knots and Everything — Vol. 61

# Numeral Systems with Irrational Bases for Mission-Critical Applications

**Alexey Stakhov**
International Club of the Golden Section, Canada

NEW JERSEY • LONDON • SINGAPORE • BEIJING • SHANGHAI • HONG KONG • TAIPEI • CHENNAI • TOKYO

*Published by*

World Scientific Publishing Co. Pte. Ltd.
5 Toh Tuck Link, Singapore 596224
*USA office:* 27 Warren Street, Suite 401-402, Hackensack, NJ 07601
*UK office:* 57 Shelton Street, Covent Garden, London WC2H 9HE

**Library of Congress Cataloging-in-Publication Data**
Names: Stakhov, A. P. (Alexey Petrovich), author.
Title: Numeral systems with irrational bases for mission-critical applications /
Alexey Stakhov (International Club of the Golden Section, Canada).
Description: New Jersey : World Scientific, 2018. | Series: Series on knots and everything ; vol. 61 |
Includes bibliographical references and index.
Identifiers: LCCN 2017030467 | ISBN 9789813228610 (hardcover : alk. paper)
Subjects: LCSH: Golden section. | Fibonacci numbers.
Classification: LCC QA466 .S7825 2018 | DDC 512.7/2--dc23
LC record available at https://lccn.loc.gov/2017030467

**British Library Cataloguing-in-Publication Data**
A catalogue record for this book is available from the British Library.



Printed in Singapore

# Contents

# Preface

## 1. The most important applied problems, which stimulated the development of mathematics on the stage of its origin

The great Russian geometer Nikolay Lobachevsky begins his famous "*Geometric study on the theory of parallel lines*" by the following words:

"*In geometry, I found some imperfections, which, in my opinion, are the reason why this science until now is in the state, in which it had come to us from Euclid. I attribute to these imperfections the following: the vagueness in the first definitions of geometric quantities, the methods of measurement of these quantities, and, finally, the important gap in the theory of parallel lines.*"

It is important to emphasize that in this quote Lobachevsky posed the *problem of the methods for measurement of quantities* on one par with the problem of creation of the *parallel lines theory*, solution of which brought Lobachevsky the world glory as the creator of hyperbolic geometry; according to academician Andrey Kolmogorov [1], Lobachevsky's geometry is the most important mathematical discovery of the 19th century.

By discussing the process of origin of mathematics, Kolmogorov points out in [1] two applied problems, which stimulated the development of mathematics at the stage of its origin:

"*Counting of items on the very early stages of development of culture has led to the creation of the elementary concepts of arithmetic of natural numbers. Only on the basis of the oral numeration systems, the written numeral systems arose, and also gradually the methods of implementation of the four arithmetic operations over natural numbers were worked out...*

*Measurement tasks (the amount of grain, road length, etc.) had led to the emergence of names and designations of the simplest fractional numbers and to the development of the methods for performing arithmetic operations in fractional numbers. Thus, it was accumulated material that had led gradually to arithmetic, the most ancient branch of mathematics. Measurement of areas and volumes, demands of construction machinery, and a little later - astronomy, caused the development of the initial notions of geometry.*"

In this quote Kolmogorov in compressed form had reflected the development of mathematics at the stage of its origin, and the role of *counting* and *measurement* in the development of arithmetic and geometry, the oldest branches of mathematics.

Similar idea had been developed in the article [2], written by the famous Russian historians of mathematics Bashmakova J.G. and Youshkevich A.P. (Moscow University).

## 2. Johannes Kepler and Alexey Losev about the "golden section" as the "golden" paradigm of ancient Greek science

As it is known, in ancient Greek philosophy, Harmony was the opposite to Chaos. Harmony means the self-organization of the Universe. Alexey Losev, the famous Russian philosopher of the aesthetics of antiquity and the Renaissance, evaluates the main achievements of the ancient Greeks in this area as follows [3]:

"*From Plato's point of view, and in general in terms of the entire ancient cosmology, the Universe was determined as a certain proportional whole, which obeys to the law of harmonic division - the golden section .... The ancient Greek system of cosmic proportions in the literature is often interpreted as a curious result of unrestrained and wild imagination. In such explanations we see the scientific helplessness of those who claim this. However, we can understand this historical and aesthetic phenomenon only in connection with a holistic understanding of history, that is, by using a dialectical view on culture and by searching for the answer in the peculiarities of ancient social life.*"

Here Losev formulates the "golden" paradigm of ancient cosmology. It was based upon the most fundamental ideas of ancient science that are sometimes treated in modern science as a "*curious result of an unrestrained and wild imagination.*" First of all, we are talking about the *Pythagorean Doctrine of the Numerical Harmony of the Universe* and *Plato's cosmology*, based on the *Platonic solids*. By referring to the geometrical structure of the Cosmos and its mathematical relations, expressing Cosmic Harmony, the Pythagoreans had anticipated the modern mathematical basis of the natural sciences, which began to develop rapidly in the 20th century. Pythagoras's and Plato's ideas about Cosmic Harmony proved to be immortal.

Johannes Kepler, brilliant astronomer and the author of "Kepler's laws," expressed his admiration of the golden ratio in the following words:

"*Geometry has two great treasures: the first of them is the Theorem of Pythagoras; the second of them is the division of a line into extreme and mean ratio. The first one we may compare to a measure of gold; the second we may name a precious stone*" (taken from [4]). You should recall that the ancient problem of the division of a line segment in extreme and mean ratio is Euclid's language for the golden section!

Thus, the idea of Harmony, which underlies the ancient Greek doctrine of Nature, was the main "paradigm" of Greek science, starting from Pythagoras and ending by Euclid. This paradigm relates directly to the golden section and Platonic solids, which are the most important Greek geometric discoveries for the expression of the Universe Harmony.

## 3. Pythagoreanism and Pythagorean MATHEM's

By studying sources of mathematics origin, we inevitably come to Pythagoras and his doctrine, named *Pythagoreanism* [5]. According to the tradition, *Pythagoreans* were divided into two separate schools of thought, *mathēmatikoi* (*mathematicians*) and *akousmatikoi* (*listeners*). *Listeners* had developing religious and ritual aspects of *Pythagoreanism*, *mathematicians* studied four Pythagorean MATHEM's: *arithmetic, geometry, harmonics* and *spherics*. These MATHEM's, according to Pythagoras, were the main components of mathematics. Unfortunately, the Pythagorean MATHEM of *harmonics* was lost in mathematics in the process of its historical development.

## 4. The most important mathematical discovery on the stage of the origin of mathematics

In the period of the mathematics origin, one of the "key" mathematical discoveries, influenced on the further development of mathematics and science in general, had been made. We are talking about the positional principle of numbers representation. It is emphasized in [2] that "*the Babylonian numeral system with the base of 60 what arose approximately in 2000 BC, was the first numeral system, based on the positional principle*".

It is necessary to note that the *positional principle of number representation* and *positional numeral systems*, in particular, decimal and binary systems, became one of the "key" ideas of mathematical education and modern computer science.

Unfortunately, historians of computer science, who are familiar with the developed computer science theory, sometimes forget about the role, which was played by the numeral systems in the history of computers. The first realizations of the simplest calculating devices (abacus and adding machines), the prototypes of today's computers, appeared long before the emergence of the Boolean algebra, theory of algorithms and other important theories. At that, namely the numeral systems and the rules of the simplest arithmetical operations had played a major role in the creation of these simplest calculating devices. It should not be forgotten about this, when we predict further development of computer science. The numeral systems have to play a crucial role in future development of modern computer science.

Why is it that in number theory and theoretical arithmetic the numeral systems weren't attract such attention, which they undoubtedly had deserved? The whole point is in "tradition." In Greek mathematics, which reached a high level of development, for the first time there appeared the separation of mathematics into two parts, the "*higher*" *mathematics*, which included *geometry* and *number theory*, and "*logistics,*" the applied science about the art of arithmetical calculations ("school" arithmetic), geometric measuring and constructing.

Already since Plato's time, "logistics" was treated as applied discipline, which wasn't included in education circle of philosopher and scientist. The ascending to Plato dismissive attitude towards the "school" arithmetic and its problems, and the lack of a sufficiently serious need for the creation of new numeral systems in computing practice, which during the last centuries was entirely satisfied with the *decimal system*, and during the last decades with the binary system (in computer science), may explain the fact, why in the theory of numbers there existed the neglectful attitude to numeral systems and why in this part number theory isn't much moved forward in comparison with the period of the origin of mathematics.

## 5. Researches in the field of numeral systems in modern computer science

The situation had sharply changed at the second half of the $20^{th}$ centuries after the occurrence of modern computers. In this area a huge interest in methods of number representation and new computer arithmetic's was again arisen. During this period the numeral systems with the "exotic" titles and properties appeared: *system for residual classes, ternary symmetrical numeral system, numeral system with the complex radix, nega-positional, factorial, binomial numeral systems* [6], etc. All of them had those or other advantages in comparison with the binary system and were directed on the improvement of those or other computer characteristics; some of the new numeral systems became the basis for the creation of new computer projects (the ternary computer "Setun," the computer, based on system for residual classes and so on).

But there is also other interesting aspect of this problem. Later 4 millennia after the invention of Babylonian positional principle of number representation, we can see a peculiar "Renaissance" in the field of numeral systems [6]. Due to the efforts first of all of the experts in computer science, *mathematics as though again turned back to the period of its origin*, when the numeral systems had defined a topic and essence of all mathematics (Babylon, Ancient Egypt, India).

In fact, in the opinion of many historians of mathematics, this period was considered extremely important for the development of mathematics

[1, 2]. For example, namely during this period a concept of *natural number*, the main concept of number theory, was developed [2].

But then we can ask the following question: possibly the modern numeral systems, created for computer needs, could affect the development of number theory and in such way could affect not only in the development of computer science, but also of all mathematics. A search of the answer to this question is the major goal of Chapters 3 and 4, devoted to *Bergman system* and their generalizations, *Fibonacci p-codes, codes of the golden p-proportions*, and also *ternary mirror-symmetrical arithmetic, Fibonacci and "Golden" arithmetic's*, which underlie the concept of *Fibonacci computers*.

## 6. Mission-critical applications

The computer science and digital metrology are at a new stage of their development, at the stage of designing computing and measuring systems for **mission-critical applications.**

In the Wikipedia article "Mission critical" [7] we read:

"***Mission critical*** *refers to any factor of a system (components, equipment, Personnel, process, procedure, software, etc.) that is essential to business operation or to an organization. Failure or disruption of* ***mission critical*** *factors will result in serious impact on* ***business operations*** *or upon an organization, and even can cause social turmoil and catastrophes. Therefore, it is extremely critical to the organization's "mission" (to avoid Mission Critical Failures).*

***Mission critical system*** *is a system whose failure may result in the failure of some goal-directed activity. Mission essential equipment and mission critical application are also known as mission-critical system. Examples of mission critical systems are: an online banking system, railway/aircraft operating and control systems, electric power systems, and many other computer systems that will adversely affect business and society seriously if downed.*

*A good example of a mission critical system is a navigational system for a spacecraft. The difference between mission critical and business critical lies in the major adverse impact and the very real possibilities of loss of life, serious injury and/or financial loss. A business-critical system fault can influence a single company or several, and can partially stop lifetime activity (hours or days).*"

The famous Russian expert in computer science academician Khetagurov in one of his articles [8] discusses the problem of the use of modern microprocessors, based on the binary system, in terms of national security:

"*The use of microprocessors, controllers, and software computing resources of a foreign proceeding to solve problems in real-time systems of military, administrative and financial destination is fraught with big problems. This is a sort of "Trojan horse", a role of which only now became manifest. Losses and damage from their use can significantly affect the national security of Russia...*"

Thus, academician Khetagurov raises the question of the creation of modern computational tools, having built-in system of error detection for ensuring high informational reliability and noise-immunity of the mission-critical systems. This problem is not new, but its solution is far from its completion due to the lack of sufficient effective scientific solutions in this area.

Unfortunately, the following not very optimistic conclusions follow from the above-mentioned arguments of academician Khetagurov:

1. For the case of *mission-critical systems,* mankind has become a hostage of the **binary system**, which has a "zero redundancy" and does not contain internal mechanisms for detecting errors in the computer and measuring systems. The further usage of the binary system in microprocessors, microcontrollers and informational technology should be declared inadmissible for *mission-critical applications.*
2. All previous experience of improving the noise-immunity of informational systems cannot be used for effective solution of all problems of mission-critical systems. This concerns, in particular, the theory of error-correcting codes [9], which was created to improve the noise-immunity of **SERIAL** data transmission systems. Therefore, most error-correcting codes (cyclic codes, Reed-Solomon codes, etc.) cannot be effectively used to protect modern computer structures, in particular, electronic memory, in which data are represented in **PARALLEL** form.

3. The *Hamming code*, widely used to protect electronic memory, has, in fact, a very significant **drawback**, from the points of view *mission-critical applications*. It guarantees 100% correction of single-bit errors, but at the same time performs "false" correction of all the odd many-bit errors of big multiplicity (3, 5, 7, etc.). This dramatically reduces the error-detection ability of the Hamming code. The effect of "false" correction of errors of odd many-bit multiplicity leads to the appearance of the "false" output data which is unacceptable for *mission-critical systems* and can lead to catastrophes.
4. Since most of the existing *error-correcting codes* are **not arithmetic codes**, in order to detect errors in arithmetic operations, we need to use the so-called arithmetic codes, which leads to a significant complication of computer structures and the appearance of **encoding-decoding problem**, when the coding–decoding devices are more complex than those computer devices that they protect against errors. This means that we have to protect from errors and the coding–decoding devices themselves (**encoding-decoding problem**).
5. Thus, the existing theory of designing computer and measuring systems is insufficient to protect effectively analog-to-digit and digit-to-analog structures from external and internal influences that can lead to "false" output signals of computer and measuring systems what can cause technological disasters. *This means that the mission-critical systems have identified all the shortcomings of existing methods of designing the noise-immune and stable computer and measuring systems. To overcome these shortcomings, new scientific ideas and approaches to designing the mission-critical systems are needed.*

## 7. The main ideas and goals of this book

1. The main idea is replacing the classic binary system on the binary (0,1) or ternary (−1,0,1) redundant numeral systems, which have code redundancy, enough for organization of effective detection of errors in the data, represented in **PARALLEL** form.

2. For the realization of this idea, a new surge of interest in the noise-immune positional numeral systems as the basis of new computing and measuring projects has arisen in modern computer science. Among them, *the binary (0,1) numeral system with the irrational base* $\Phi = \frac{1+\sqrt{5}}{2}$ (the golden ratio), proposed in 1957 by the American 12-year wunderkind George Bergman [10], is of special interest both for the number theory and also for computer science. Unfortunately, *Bergman's system* did not have any consequences for computer science, because computer experts of that period and Bergman himself could not appreciate the importance of Bergman's mathematical discovery for computer science.

3. The author of this book is one of the first experts in the field of computer science, who, by following Professor Donald Knuth, evaluated the importance of the numeral systems with irrational bases for designing of noise-immune computing and measuring systems. Regardless of George Bergman, the author of this book has developed new computer ideas in his Doctoral dissertation (1972) [11], which has become the source for the introduction of new classes of redundant numeral systems, the so-called ***Fibonacci p-codes*** *and* ***codes of the golden p-proportions*** for noise-immune computing and measuring systems [12–36]. These mathematical results have become a source for a wide international patenting of author's inventions in this field. 65 foreign patents of USA, Japan, France, England, Germany, Canada and other countries are legal confirmation of author's priority in this field [37–49].

4. This book is the result of many-years of author's research in this field. These results were presented in the author's two books, *Introduction into the Algorithmic Measurement Theory* (Moscow, Soviet Radio, 1977) [14], and *Codes of the Golden Proportion* (Moscow, Radio and Communications, 1984) [15]. These books were not be translated into English and are therefore not known to English-speaking audience.

5. The main goal of this book is to set forth new informational and arithmetical fundamentals of computing and measurring systems based on *Fibonacci p-codes* and *codes of the golden p-proportions*, *Bergman's system* and *"golden" ternary mirror-symmetrical arithmetic* and their

applications in number theory, computer science and digital metrology. The book is intended for a wide range of readers, including engineers and researchers in computer science and digital metrology, students of colleges and universities and their teachers in the same areas. The book is of interest to all researchers who use the *golden ratio* and *Fibonacci numbers* in their subject areas and can serve as a manual for students of colleges and universities.

The book is of historical interest, as it covers a long period in the development of the theory of positional numeral systems, beginning with the *Babylonian positional numeral system* until the latest achievements in this field.

The following basic publications underlie this book:

1. George Bergman. A number system with an irrational base [10]
2. Alexey Stakhov. Introduction into algorithmic measurement theory (Russian) [14]
3. Alexey Stakhov. Codes of the golden proportion (Russian) [15]
4. Alexey Stakhov. The mathematics of harmony. From Euclid to contemporary mathematics and computer science [4].
5. Alexey Stakhov. Brusentsov's ternary principle, Bergman's number system and ternary mirror-symmetrical arithmetic [28]

Also here the author's publications in Russian, Ukrainian, English [12, 13, 16–36], including the newest 2016 publications [33–36], and patents of USA, Japan, England, France, Germany, Canada and other countries [37–49] are used.

This book is a pioneering book in the computer science and coding theory for English-speaking audience. Essentially, we are talking about creating *new informational and arithmetical foundations of computer science and digital metrology for mission-critical applications.*

# Introduction

In connection with the development of computer science, there arose a new surge of interest in positional numeral systems, as the basis for new computer projects. Among them, two numeral systems are the most interesting: *the binary (0,1) numeral system with the irrational base* $\Phi = \frac{1+\sqrt{5}}{2} \approx 1.618$ (the *golden ratio*), proposed by the American mathematician George Bergman in 1957 [10], and ternary mirror-symmetrical numeral system with base $\Phi^2 = \frac{3+\sqrt{5}}{2} \approx 2.618$ (*square of the golden ratio*), proposed by the author of this book in 2002 article [28]. Also *Fibonacci p-codes* and *codes of the golden p-proportions* [4, 12–25, 30, 33, 35, 36] are of great theoretical interest for number theory and applied interest for computer science and digital metrology. The main feature of these new ways of positional representation of numbers is that their bases are irrational numbers of special type such as the *golden ratio* and its generalization, the *golden p-proportions*. These positional numeral systems belong to a new class of numeral systems, called *numeral systems with irrational bases* introduced by George Bergman in 1957 [10]. These numeral systems alter our ideas about positional numeral systems and can affect future development of number theory and computer science.

The main goal of the book is to set forth the fundamentals of numeral systems with irrational bases. Although this mathematical theory is of applied nature, in the course of their development this theory had gone far beyond their applications and touched the foundations of mathematics and computer science, in particular, number theory and coding theory.

The book is also of historical interest as it covers a long period in the development of the theory of positional numeral systems, beginning with the Babylonian positional numeral system [2] until the latest achievements in this field [33, 35, 36].

This book is a development and continuation of the author's first books "*Introduction to Algorithmic Measurement Theory*" (1977) [14], "*Codes of the Golden Proportion*" (1984) [15], written in Russian, and the author's 2009 book "*The Mathematics of Harmony. From Euclid to Contemporary Mathematics and Computer Science*" (2009) [4], written in English.

This book consists of 5 chapters:

Chapter 1. Preliminary Historical and Mathematical Information

Chapter 2. A New View on Numeral Systems: Unusual Hypotheses, Surprising Properties and Applications

Chapter 3. Bergman's System, "Golden" Number Theory and Mirror-Symmetrical Arithmetic

Chapter 4. Fibonacci *p*-Codes and Conception of Fibonacci Computers

Chapter 5. Codes of the Golden *p*-Proportions and Their Applications in Computer Science and "Golden" Metrology

The following historical and mathematical concepts are discussed in Chapter 1: problem of harmony, golden ratio, Platonic solids, Proclus hypothesis, Fibonacci and Lucas numbers, Fibonacci and "golden" $Q$-matrices, generalized Fibonacci $Q_p$-matrices, Fibonacci λ-numbers and so on. Most of these concepts such as harmony, the classic golden section and Fibonacci and Lucas numbers are widely known to the scientific community, but some of them, such as Proclus hypothesis, the Fibonacci and Lucas *p*-numbers, the generalized Cassini formula, the Fibonacci $Q_p$-matrices, are new mathematical results, which are not very well known to mathematicians and experts in computer science.

Chapter 2 is devoted to the description of new view on the positional numeral systems, starting from the *Babylonian numeral system with base 60* and ending with the *ternary principle of Nikolay Brusentsov*, Principal Designer of the ternary computer "Setun" (Moscow University). Here two new hypotheses about the origin of the *Babylonian positional numeral system with base 60* and *Mayan's positional numeral system with base 20* are the most interesting historical results. It is showed in Chapter 2 that both hypotheses are based on the deep connection between *Egyptian calendar* and *Dodecahedron* (*Babylonian numeral system*) and *Mayan's calendar* and *Icosahedron* (*Mayan's numeral system*).

Chapter 3 is one of the most important chapters of the book. It is based on the mathematical discovery of the young American mathematician George Bergman, who at the age of 12 introduced the first (in world history) *positional numeral system with irrational base* $\Phi = \frac{1+\sqrt{5}}{2} \approx 1.618$ (the *golden ratio*) [10]. Unfortunately, in 1957 this numeral system wasn't evaluated properly by mathematicians and experts in computer science, despite the fact that it is the profound mathematical idea, which can lead to revolutionary changes in the field of number theory and computer science. New number theory, called the *"golden" number theory*, is described in Chapter 3. New properties of natural numbers, based on *Bergman's system*, are one of the most important results of Chapter 3.

The greatest theoretical and practical interest to experts in the field of coding theory and computer science is of the *"golden" ternary mirror-symmetrical numeral system with base* $\Phi^2 = \frac{3+\sqrt{5}}{2} \approx 2.618$ [28]; it can be used to create a unique "golden" ternary processors and computers. The outstanding American mathematician and a world expert in the field of computer science Professor Donald Knuth was the first prominent scientist, who congratulated the author with the publication of the article [28].

Chapter 4 describes a new class of positional numeral systems with irrational bases, *the Fibonacci p-codes* and new computer arithmetic, named *Fibonacci arithmetic*. These scientific results can be used to design Fibonacci computers, a new class of noise-immune computers,

Chapter 5 describes a new class of positional numeral systems with irrational bases, the *codes of the golden p-proportions and their applications in computer science and digital metrology*. The self-correcting analog-to-digit and digit-to-analog converters are the main result of this chapter.

The book is written in popular form and intended for a wide range of readers, including students of colleges and universities in the field of mathematics, computer science and digital metrology, and also for teachers of colleges and universities in the same areas. The book is of interest to all researchers who use the golden ratio and Fibonacci numbers in their subject areas.

# Acknowledgements

About 50 years ago the author had read the remarkable brochure *Fibonacci Numbers* written by the famous Soviet mathematician Nikolay Vorobyov. This brochure was the first mathematical work on *Fibonacci numbers* published in the second half of the 20th century. This brochure determined author's scientific interests in Fibonacci numbers. In 1974 the author met with Professor Vorobyov in Leningrad (now St. Petersburg) and discussed with him the author's scientific achievements in this area. He gave the author as a keepsake his brochure *Fibonacci Numbers* with the following inscription: *"To highly respected Alexey Stakhov with Fibonacci's greetings."*

The author's expresses great thanks to his teacher, the outstanding Ukrainian scientist, Professor Alexander Volkov; under his scientific leadership the author defended PhD dissertation (1966) and then DrSci dissertation (1972). These dissertations were the first step in the author's research, which led the author to the conceptions of Mathematics of Harmony and Fibonacci computers, based on the *golden ratio* and *Fibonacci numbers*.

In the stormy scientific life, the author met many fine people, who could understand and evaluated the author's enthusiasm and appreciate his scientific direction. With deep gratitude, the author recollects a meeting with the famous Austrian mathematician Alexander Aigner (Graz, Austria, 1976). The meeting with Prof. Aigner was the beginning of the international recognition of the author's scientific direction. Another remarkable person, who had a great influence on the author's research was the Ukrainian mathematician Yuri Mitropolskiy. Thanks to the support of Yury Mitropolskiy, the author had published many important articles in various Ukrainian academic journals.

Author's arrival to Canada in 2004 became the beginning of new stage in author's scientific research. Within 13 years, the author has published 50 fundamental articles in different international English-language journals. The publication of the two fundamental books *The Mathematics of Harmony* (World Scientific, 2009) and *The "Golden" Non-Euclidean Geometry* (World Scientific, 2016) is the author main scientific achievements of the Canadian period. These books were published, thanks to the support of the famous American mathematician Prof. Louis Kauffman, editor of the Series on Knots and Everything (World Scientific) and Prof. M.S. Wong, the famous Canadian mathematician (York University) and editor of the Series on Analysis, Application and Computation (World Scientific).

This book has been published by the initiative of Prof. Louis Kauffman. A huge help in editing of the above two books was rendered by the American philosopher Prof. Scott Olsen, one of the leading American experts in the field of *golden section*. The author expresses deep gratitude to these scientists for the support.

Lastly, this book would never have been written without self-denying support of my wife Antonina, who always created the perfect conditions for scientific work in any countries, where the author worked. She had been sailing together with the author for more than 50 years on the "Golden" journey to different countries and continents (Europe, Africa (Libya and Mozambique), America (Canada)). In addition, the author would like to express his special thanks to his daughter Anna Sluchenkova for her critical remarks, and her invaluable help in the English translation and editing of the book, and, especially, for her work in preparing illustrations, and coordination and final preparation of camera-ready manuscript. Without her support this book was never been published.

Alexey Stakhov

## Chapter 1

# Preliminary Historical and Mathematical Information

## 1.1. The idea of harmony in its historical development

### 1.1.1. *Harmonic ideas of Pythagoras, Plato and Euclid*

What is the major idea of the ancient Greek science? A majority of researchers give the following answer: the idea of Harmony associated with the *golden ratio* and *Platonic Solids*. Pythagoras, Plato and Euclid were the most outstanding representatives of this trend in the ancient Greek science, philosophy and mathematics. The greatest interest to the concept of harmony, that is, to the ideas of Pythagoras, Plato and Euclid, always arose in the periods of greatest prosperity of the "human spirit." From this point of view, in the studying of the Mathematics of Harmony [4] we can highlight the following critical periods.

### 1.1.2. *The ancient Greek period*

Conventionally, it can be assumed that this period starts with the research of Pythagoras and Plato. Euclid' *Elements* became a final event of this important period. According to Proclus' hypothesis [34], Euclid created his *Elements* in order to create the complete geometric theory of the five Platonic solids, which have been associated in the ancient Greek science with the Universe Harmony. The geometric fundamentals of the theory of Platonic solids (Fig. 1.1) were described by Euclid in the concluding Book (Book XIII) of the *Elements*.

In addition, Euclid simultaneously introduced in *Elements* some advanced achievements of ancient Greek mathematics, in particular, the *golden section* (Book II), which was used by Euclid for the creation of the geometric theory of the *Platonic solids* (Book XIII).

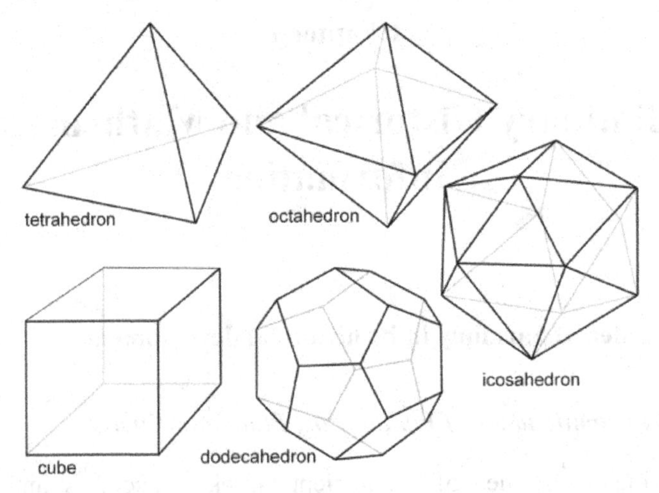


Figure 1.1. Platonic solids: tetrahedron, octahedron, cube, dodecahedron, icosahedron.

### 1.1.3. *The Middle Ages*

In the Middle Ages, very important mathematical discovery was made. The famous Italian mathematician Leonardo of Pisa (Fibonacci) wrote the book "Liber Abaci" (1202). In this book, he described *the task of rabbits reproduction*. By solving this problem, he found the remarkable numerical sequence — the Fibonacci numbers $F_n$:

$$1,1,2,3,5,8,13,21,34,55,89,\ldots, \tag{1.1}$$

which are given by the recurrent relation:

$$F_n = F_{n-1} + F_{n-2};\ F_1 = F_2 = 1. \tag{1.2}$$

### 1.1.4. *The Renaissance*

This period is connected with the names of the prominent figures of the Renaissance: Piero della Francesca (1412–1492), Leon Battista Alberti (1404–1472), Leonardo da Vinci (1452–1519), Luca Pacioli (1445–1517), Johannes Kepler (1571–1630). In that period three books, which were the best reflection of the idea of the "Universe Harmony," were published. The first of them is the book *Divina Proportione* ("The Divine

Proportion") (1509). This book had been written by the outstanding Italian mathematician and scholar monk Luca Pacioli under the direct influence of Leonardo da Vinci, who illustrated Pacioli's book.

Also the brilliant astronomer of 17th century Johannes Kepler made an enormous contribution to the development of the "harmonic ideas" of Pythagoras, Plato and Euclid.

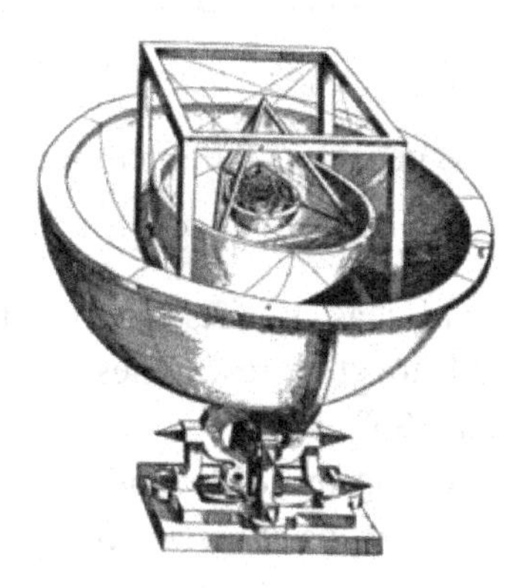

Figure 1.2. Kepler's Cosmic Cup.

In his first book *Mysterium Cosmographicum* (1596) he built the so-called *Cosmic Cup* (Fig. 1.2), the original model of the Solar system, based on the Platonic solids. The book *Harmonice Mundi* (Harmony of the World) (1619) is the main Kepler's contribution into the Doctrine of the Universe Harmony. In the *Harmony*, he attempted to explain the proportions of the Universe — particularly the astronomical and astrological aspects — by using musical terms. The *Musica Universalis* or *Music of the Spheres*, which had been studied by Pythagoras, Ptolemy, was the central idea of Kepler's *Harmony*.

### 1.1.5. *The 19th century*

This period is connected with the works of the French mathematicians Jacques Philippe Marie Binet (1786–1856), Francois Edouard Anatole Lucas (1842–1891), German poet and philosopher Adolf Zeising (1810–1876) and the German mathematician Felix Klein (1849–1925).

Jacques Philippe Marie Binet derived a mathematical formula (*Binet's formula*) to represent the Fibonacci numbers through the "golden ratio" $\Phi = \frac{1+\sqrt{5}}{2}$ (see below).

Francois Edouard Anatole Lucas introduced the Lucas numbers $L_n$ similar to the Fibonacci numbers $F_n$ (1.1); the Lucas numbers $L_n$ are calculated by the recurrent relation similar to (1.2), but with other seeds:

$$L_n = L_{n-1} + L_{n-2}; \quad L_1 = 1, L_2 = 3. \tag{1.3}$$

The recurrent relation (1.3) generates the Lucas numbers:

$$L_n : 1, 3, 4, 7, 11, 18, 29, \ldots. \tag{1.4}$$

The merit of Binet and Lucas consists of the fact that their researches became a launching pad for the researches of American Fibonacci Association, established in 1963.

German poet Adolf Zeising in 1854 published the book *Neue Lehre von den Proportionen des menschlichen Körpers aus einem bisher unerkannt gebliebenen, die ganze Natur und Kunst durchdringenden morphologischen Grundgesetze entwickelt*. The basic idea of Zeising is to formulate the Law of proportionality. He formulated this Law as follows:

*"A division of the whole into unequal parts is proportional, when the ratio between the parts is the same as the ratio of the bigger part to the whole; this ratio is equal to the golden mean."*

The famous German mathematician Felix Klein in 1984 published the book "*Lectures on the icosahedron and the solution of equations of the fifth degree*" [50] dedicated to the geometric theory of the icosahedron and its role in the general theory of mathematics. Klein treats the icosahedron as a mathematical object, which is a source for the five mathematical theories: *geometry, Galois theory, group theory, invariant theory and differential equations*.

What is the significance of Klein's ideas from the point of view of the Mathematics of Harmony [4]? According to Klein, the Platonic icosahedron, based on the *golden ratio*, is the main geometric figure of mathematics. It follows from this that the *golden ratio*, which is the main geometric proportion of icosahedron, is the main geometric object of mathematics, which, according to Klein, can unite all the basic mathematical theories.

This Klein's idea is consistent with the ideas of author's article [33]. The article presents the concept of the "*golden*" *number theory*, based on the *golden ratio* and its generalizations.

### 1.1.6. *The first half of the 20th century*

In the first half of the 20th century the development of the "golden" paradigm of the ancient Greeks is associated with the names of the Russian Professor of architecture G. D. Grimm (1865–1942) and the classic of the Russian religious philosophy Paul Florensky (1882–1937).

In the theory of architecture, Grimm's 1935 book "Proportionality in Architecture" [51] is well-known. The purpose of the book was formulated in the "Introduction" as follows:

*"In view of the exceptional significance of the Golden Section in the sense of the proportional division, which establishes a permanent connection between the whole and its parts and gives a constant ratio between them (which is unreachable by any other division), the scheme, based on it, is the main standard and is accepted by us in the future as a basis for checking the proportionality of historical monuments and modern buildings ... By taking this general importance of the Golden Section in all aspects of architectural thought, the theory of proportionality, based on the proportional division of the whole into parts, corresponding to the Golden Section, should be recognized as the architectural basis of proportionality at all."*

In 1920, Paul Florensky wrote the work "*At the watershed of a thought.*" Its third chapter is devoted to the "golden ratio". The Belorussian philosopher Edward Soroko in the book [52] evaluates Florensky's work as follows:

*"The aim was to derive analytically the stability of the whole object, which is in the field of the effect of oppositely oriented forces. The project was conceived as an attempt to use the "golden ratio" and its substantial basis, which manifests itself not only in a series of experimental observations of nature, but on the deeper levels of*

*knowledge, for the case of penetration into the dialectic of movement, into the substance of things."*

### 1.1.7. *The second half of the 20th century and the 21st century*

In the second half of 20th century the interest in this area is increasing in all areas of science, including mathematics. The Soviet mathematician Nikolay Vorobyov (1925–1995) [53], the American mathematician Verner Hoggatt (1921–1981) [54], the American mathematician Thomas Koshy [55] and others became the most outstanding representatives of this direction in mathematics.

The revival of the idea of harmony in modern science is determined by new scientific realities. The penetration of the Platonic solids, the golden ratio and Fibonacci numbers into all areas of theoretical natural sciences (crystallography, chemistry, astronomy, earth science, quantum physics, botany, biology, geology, medicine, genetics, etc.), as well as into computer science and economics was the main reason for the renewed interest in the ancient idea of the Universe Harmony in modern science and the stimulus for the development of the Mathematics of Harmony [4].

## 1.2. Proclus hypothesis: A new view on Euclid's *"Elements"* and the history of mathematics

### 1.2.1. *Proclus hypothesis*

"Proclus hypothesis" [34], formulated in the 5th century AD by the famous Greek philosopher Proclus Diadochus (412–485), who was one of the best commentators of Euclid's *Elements,* contains the unexpected view on Euclid's *Elements.*

According to Proclus, Euclid's goal was not to set forth the geometry itself, but to build the complete theory of regular polyhedra ("Platonic solids"). This theory was outlined by Euclid in Book XIII, that is, the concluding book of the *Elements* what in itself is an indirect confirmation of *Proclus' hypothesis* because the most important material of scientific work, its purpose and mission are placed traditionally in its final part.

To solve this problem, Euclid had included the necessary mathematical material into the Elements. The most curious thing is that he had introduced in Book II the *golden ratio*, what is unexpected and inexplicable for many historians of mathematics. However, *Proclus' hypothesis* explains this fact very simply. Euclid used the golden ratio (*DEMR, a division in the extreme and mean ratio*) for the creation of geometric theory of regular polyhedra. In Plato's Cosmology, the regular polyhedra has been associated with the Universe Harmony. This means, that Euclid's *Elements* are based on the "harmonic ideas" of Pythagoras and Plato, that is, Euclid's *Elements* are historically the first variant of the *Mathematics of Harmony*. This unexpected view on the *Elements* leads to the conclusion, which changes our view on the history and structure of mathematics.

### 1.2.2. *Again about Kolmogorov's two problems of mathematics on the stage of its origin*

As it was mentioned above, Kolmogorov in the book [1] had identified the two main, that is, "key" problems, which stimulated the development of mathematics at the stage of its origins, the *counting problem* and the *measurement problem*. However, it follows from *Proclus' hypothesis* [34] another "key" problem: the *harmony problem*, which underlies Euclid's *Elements*.

Thus, we can see that there are three "key" problems, the *counting problem*, the *measurement problem*, and the *harmony problem*, which stimulated development of mathematics on the stage of its origin (see Fig. 1.3).

The first two "key" problems resulted in the creation of two fundamental notions of mathematics — *natural numbers* and *irrational numbers*, which underlie the *Classic Mathematics*. The harmony problem, connected with the *golden ratio* or *DEMR* (Proposition II.11 of Euclid's *Elements*) resulted in the *Harmony Mathematics* [4], the new inter-disciplinary direction of contemporary science and mathematics.

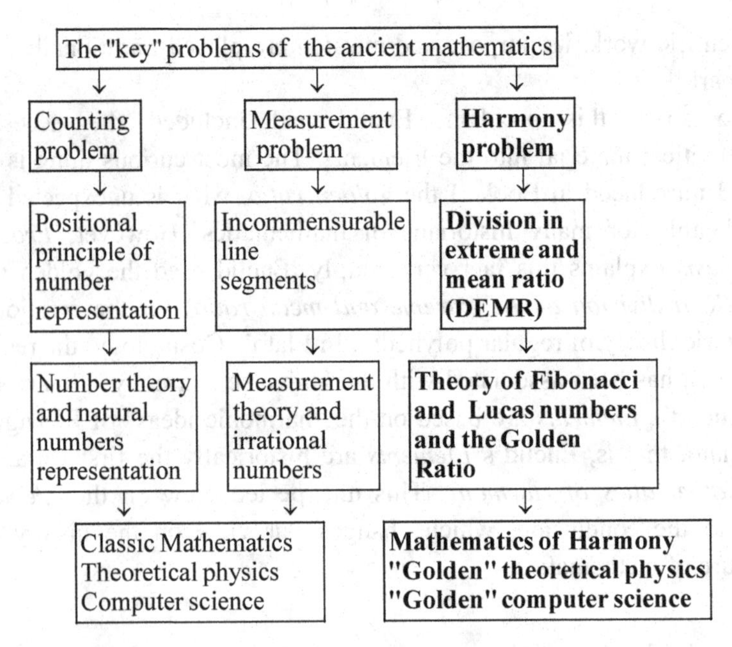


Figure 1.3. Three "key" problems of the ancient mathematics.

This approach leads to the unexpected conclusion. It turns out, that in parallel with the *Classic Mathematics*, another direction, the *Mathematics of Harmony*, was developing in mathematics starting since ancient time.

Similarly to the *Classic Mathematics*, the *Mathematics of Harmony* takes its origin in Euclid's *Elements*. However, the Classic Mathematics focuses on the axiomatic approach and other ancient achievements (number theory, theory of irrationalities and so on), while the Mathematics of Harmony is based on the *golden ratio* or *DEMR* (Proposition II.11) and *Platonic solids*, described in Book XIII of Euclid's *Elements*. Thus, Euclid's *Elements* is a source of two independent mathematical directions: the *Classic Mathematics* and the *Mathematics of Harmony*.

For many centuries, the creation of the *Classic Mathematics*, the *Czarina of Natural Sciences*, was the main focus of mathematicians. However, starting from Pythagoras, Plato, Euclid, Pacioli, Kepler, the intellectual forces of many prominent mathematicians and thinkers were directed towards the development of the basic concepts and applications of the *Mathematics of Harmony*. We have no right to negate this important fact in the history of mathematics.

Unfortunately, these two important mathematical directions (Classic Mathematics and Mathematics of Harmony) evolved separately from one another. These important mathematical directions should be united.

A new approach to the origin of mathematics (see Fig. 1.3) is very important for the school mathematical education. This approach introduces, in a very natural manner, the idea of harmony and the golden ratio into school mathematical education. This gives the students the access to ancient science and to its main achievement — the harmonic ideas — and to tell them about the most important architectural and sculptural works of the ancient art, based upon the golden ratio (including pyramid of Khufu (Cheops), Nefertiti, Parthenon, Doryphoros, Venus and so on).

### 1.2.3. *A discussion of Proclus' hypothesis in the historical-mathematical literature*

The analysis of Proclus' hypothesis is found in many mathematical sources. Consider some of them [56–58]. In the book [56] we read: "*According to Proclus, the main objective of the "Elements" was to present the geometric construction of the so-called Platonic solids.*"

In the book [57], this idea got a further development: "*Proclus, by mentioning all previous mathematicians of Plato's circle, said: "Euclid lived later than the mathematicians of Plato's circle, but earlier than Eratosthenes and Archimedes ... He belonged to Plato's school and was well acquainted with Plato's philosophy and his cosmology; that's why he put a creation of the geometric theory of the so-called Platonic solids as the main purpose of the Elements.*"

This comment is very important and draws our attention to the deep connection of Euclid with Plato. Euclid fully shared Plato's philosophy and cosmology, based on Platonic solids, that is why, Euclid puts forward the creation of the geometric theory of Platonic solids as the main purpose of the *Elements*.

In the book [58], there is discussion on the influence of Plato and Euclid's ideas on Johannes Kepler at designing the Kepler's *Cosmic Cup* (Fig. 1.2) in his first book *Mysterium Cosmographicum*:

*"Kepler's project in Mysterium Cosmographicum was to give "true and perfect reasons for the numbers, quantities, and periodic motions of celestial orbits." The perfect reasons must be based on the simple mathematical principles, which had been found by Kepler in the Solar system, by using multiple geometric demonstrations. The general scheme of his model was borrowed by Kepler from Plato's Timaeus, but the mathematical relations for the Platonic solids (pyramid, cube, octahedron, dodecahedron, icosahedron) were taken by Kepler from the works by Euclid and Ptolemy. At that, Kepler followed Proclus and believed that "the main goal of Euclid was to build a geometric theory of the so-called Platonic solids." Kepler was completely fascinated by Proclus, he often quotes him and calls him "Pythagorean."*

From this quote, we can conclude that Kepler used the Platonic solids to create the *Cosmic Cup*, but all the mathematical relations for the Platonic solids were borrowed by him from Book XIII of the *Elements*, that is, he united in his studies Plato's Cosmology with Euclid's *Elements* system (Fig. 1.2).

## 1.3. The golden ratio in Euclid's *Elements*

### 1.3.1. *Proposition II.11*

In Euclid's *Elements* we have the concept, which later played significant role in the development of science. This concept is called the *division in extreme and mean ratio* [DEMR]. In the *Elements* this concept occurs in two forms. The first was formulated as Proposition 11 of Book II.

**Proposition II.11.** *Divide a given line segment* $AD$ *into two unequal parts* $AF$ *and* $FD$ *so that the area of the square, which is built on the larger segment* $AF$ *would be equal to the area of the rectangle, which is built on the segment* $AD$ *and the smaller segment* $FD$.

Depict this problem geometrically (Fig. 1.4).

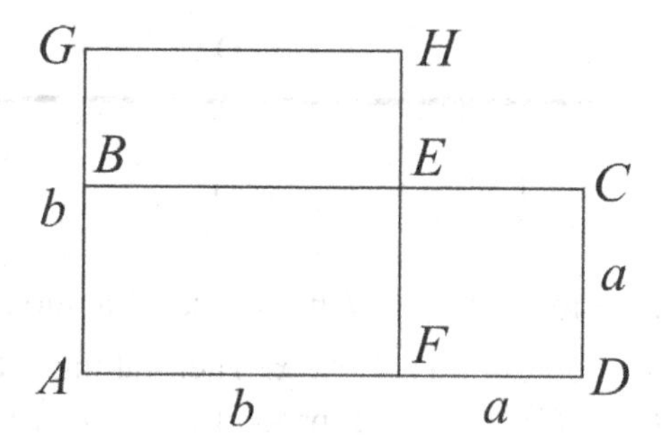


Figure 1.4. A division of a line segment in extreme and mean ratio (the "golden ratio").

Thus, according to Proposition II.11, we should choose the point *F* on the line segment *AD* so that the area of the square *AGHF* was equal to the area of the rectangle *ABCD*. If we denote the length of the larger segment *AF* of *AD* through *b* (it is equal to the side of the square *AGHF*), and the side of the smaller segment *FD* through *a* (it is equal to the vertical side *CD* of the rectangle *FECD*), then the condition for Proposition II.11 can be written as follows:

$$b^2 = a \times (a + b). \tag{1.5}$$

In Euclid's *Elements*, we have another form of the "golden ratio." This form follows from the first form, given by (1.5), if we make the following simple transformations. By dividing both parts of (1.5) first by *a*, and then by *b*, we get the following proportion:

$$\frac{b}{a} = \frac{a+b}{b}. \tag{1.6}$$

The proportion (1.6) has the following simple geometric interpretation (Fig. 1.5). Divide the line segment *AB* at the point *C* such that the larger part *CB* is related to the smaller part *AC* in the same way that segment *AB* is related to its larger part *CB* (Fig. 1.5), that is,

$$\frac{AB}{CB} = \frac{CB}{AC}. \tag{1.7}$$

Figure.1.5. The golden ratio.

This is the definition of the *golden ratio*, used in modern science.

We depict the proportion (1.7) by $x$. Then taking into consideration the fact that $AB = AC + CB$, the proportion (1.7) can be written as follows:

$$x = \frac{AC + CB}{CB} = 1 + \frac{AC}{CB} = 1 + \frac{1}{\frac{CB}{AC}} = 1 + \frac{1}{x}. \tag{1.8}$$

The following algebraic equation is derived from (1.8):

$$x^2 - x - 1 = 0. \tag{1.9}$$

Applications of the proportion (1.7) in real world imply that we should use the positive root of equation (1.9). We name this root the *golden ratio* and denote it by Φ:

$$\Phi = \frac{1 + \sqrt{5}}{2}. \tag{1.10}$$

### 1.3.2. *Platonic solids or how Euclid used the golden ratio?*

The question arises: why had Euclid introduced different forms of the golden ratio in the *Elements*; we can find these forms in Books II, VI and XIII? To answer this question, we return to the Platonic solids (Fig. 1.1), which was described by Euclid in Book XIII.

We begin our consideration with the simplest regular polyhedron, the *tetrahedron*, whose faces are equilateral triangles. One of the key observations is the fact that the sum of the interior angles of the polygons meeting at every vertex is always less than 360°. In the *tetrahedron* (Fig. 1.1), three equilateral triangles converge at one vertex (the sum of their interior angles is equal to $180° = 3 \times 60°$); thus, their common base forms a new equilateral triangle. The *tetrahedron* has the least number of faces amongst the Platonic Solids and is the three-dimensional analog to the simplest two-dimensional regular polygon, the *equilateral triangle*. And it has the least number of edges among the regular polygons.

The next geometric solid, formed by equilateral triangles, is *octahedron* (Fig. 1.1). In the octahedron four equilateral triangles converge at one vertex (the sum of their interior angles is equal to $240° = 4 \times 60°$); as a result we get a pyramid with a square base. If we connect two such pyramids by their bases, then the polyhedron with eight triangular faces, the *octahedron*, emerges.

Next we can connect 5 equilateral triangles (the sum of the interior angles is equal to $300° = 5 \times 60°$) at one vertex. Combining four of these will result in a polyhedron with 20 triangular faces, named *icosahedron* (Fig. 1.1).

A *square* is the next regular polygon (whose interior angle is equal to 90°). If we unite 3 squares at one vertex (the sum of their interior angles is equal to $270° = 3 \times 90°$) and then we add three new squares, we obtain a perfect polyhedron with 6 faces called a *hexahedron* or *cube* (Fig. 1.1).

Finally, there is one more opportunity to construct a regular polyhedron, based on the *regular pentagon* with an interior angle of 108°. If we connect 12 pentagons so that 3 regular pentagons converge in each vertex (the sum of their interior angles is equal to $324° = 3 \times 108°$), then we get the last Platonic solid, the *dodecahedron* (Fig. 1.1).

A *hexagon* is the next regular polygon after the *pentagon*. A hexagon has an interior angle of 120°. If we connect 3 hexagons at one vertex, we get a plane because the sum of their interior angles equals $360° = 3 \times 120°$. This means that it is impossible to construct a spatial geometric figure from the hexagons. Other regular polygons with more than six edges have interior angles greater than 120°. This means that we cannot form spatial geometric figures from them. This consideration led Euclid to conclude that there are only 5 convex regular polyhedra, with the following faces: *equilateral triangles (tetrahedron, octahedron and icosahedron), squares (cube) and pentagons (dodecahedron)*. The fact that only five convex regular polyhedra exist is surprising because there are infinite number of regular polygons!

There are surprising geometrical connections between regular polyhedra (property of *duality*). For example, *cube* (*hexahedron*) and *octahedron* are *dual* to each other. That is to say they can be obtained from one another, if the centroids of the faces of the first solid are taken as the vertices of the other and conversely. Similarly, as *cube* is dual to *octahedron*, *icosahedron* is dual to *dodecahedron*. *Tetrahedron* is dual to itself.

We can see from Fig. 1.1 that there are only three types of regular polygons, which can be the faces of the Platonic solids: the *equilateral triangle* (*tetrahedron*, *octahedron*, *icosahedron*), the *square* (*cube*) and the *regular pentagon* (*dodecahedron*). In order to construct the Platonic solids, we must first of all be able to build the two-dimensional faces of Platonic solids geometrically (i.e. with ruler and compass). For this purpose it was necessary to geometrically construct the faces of regular polyhedra: *equilateral triangle*, *square*, and *regular pentagon*. Geometrical construction of the *regular pentagon* was the greatest challenge to Euclid. Because the *regular pentagon* is based on the golden ratio, Euclid paid great attention to its study in the *Elements*, starting from Book II, where Euclid introduced this important geometric concept (Proposition II.11), and ending by Book XIII, devoted to geometric theory of *Platonic solids*.

Let us consider the *dodecahedron* and its dual *icosahedron* (Fig. 1.6).

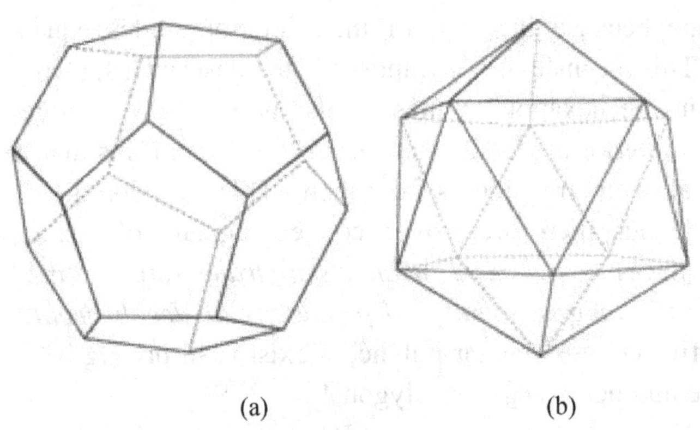

(a) (b)

Figure 1.6. Dodecahedron (a) and icosahedron (b).
Source: www.toshen.com

By using Proposition II.11, Euclid constructed the "*golden*" *isosceles triangle* (Fig. 1.7(a)). The triangle *ABD* has equal sides *AB* and *AD* and equal angles *B* and *D* at the base *BD*. These angles are equal to the doubled angle at vertex *A*.

By using the "golden" isosceles triangle *ABD* (Fig. 1.7(a)), we can construct the regular pentagon in Fig. 1.7(b). Then, only one step remains to construct the *dodecahedron* (Fig. 1.6(a)), which, according to Plato, is one of the most important regular polyhedra, symbolizing the Universe Harmony in Plato's cosmology.

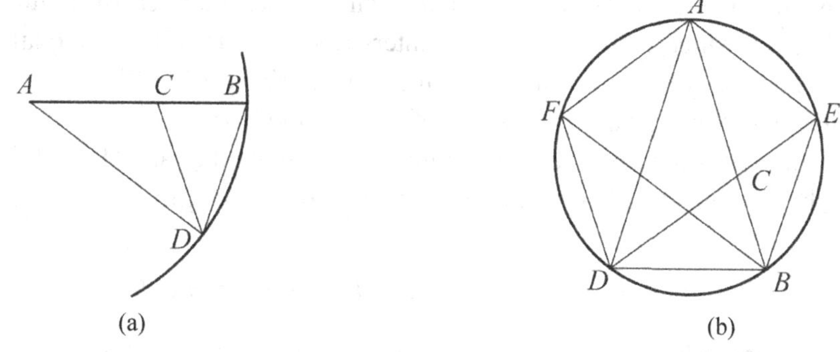


Figure 1.7. The "golden" isosceles triangle (a) and the regular pentagon (b).

Note that the regular pentagon (Fig. 1.8), which is the basis of the *dodecahedron*, was a very important geometric figure in Greek science because the *golden ratio* is the basis of the regular pentagon. The word "pentagon" is derived from the Greek word of "pentagonon."

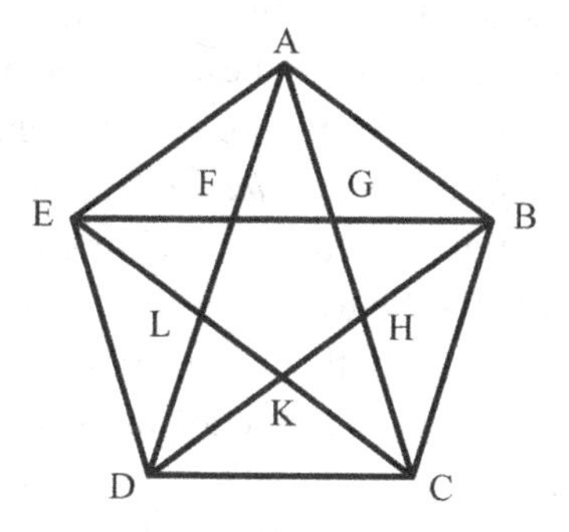


Figure 1.8. The regular pentagon and pentagram.

If we draw all diagonals in Fig. 1.8, then we get the pentagonal *star*, also known as a *pentagram* or *pentacle*. The name "pentagram" is derived from the Greek word "pentagrammon" (*pente* — five and *grammon* — line).

The points F, G, H, K, L of the intersection of the pentagon's diagonals are always golden ratio points. At the same time, they form a new regular pentagon FGHKL. In the new pentagon we may draw further diagonals. The points of their intersection produce another smaller regular pentagon. This process can be continued ad infinitum. Thus, the pentagon ABCDE contains an infinite number of regular pentagons, which are formed by the intersection points of the diagonals. This endless repetition of the same geometric figure (the regular pentagon) creates an aesthetic sense of rhythm and harmony.

The Pythagoreans were delighted with the pentagram (Fig. 1.8), which was considered as their *distinctive symbol* of recognition.

### 1.3.3. *The golden ratio in the dodecahedron and icosahedron*

The dodecahedron (Fig. 1.6(a)) and its dual icosahedron (Fig. 1.6(b)) play a special role among the Platonic solids. First of all, we emphasize that the geometry of *dodecahedron* and *icosahedron* relate directly to the *golden ratio*. Indeed, all the faces of the dodecahedron are regular pentagons, based on the *golden ratio*. If we look closely at the icosahedron (Fig. 1.6(b)), we can find that in each of its vertices there are five triangles connected together, but their outer edges form the regular pentagon. These geometric facts reveal the crucial role of the golden ratio in the geometric construction of these two Platonic solids.

There are other deep mathematical confirmations of the fundamental role of the *golden ratio* in the *dodecahedron* and *icosahedron*. For example, the Platonic solids have three specific spheres. The first (inner) sphere is inscribed inside the Platonic solid and touches the centers of its faces. We denote the radius of the inner sphere as $R_i$. The second or middle sphere of the Platonic solid touches its ribs. We denote the radius of the middle sphere as $R_m$. Finally, the third (outer) sphere is circumscribed around the Platonic solid and passes through its vertices. We denote its radius as $R_c$. In geometry, it is demonstrated that the values of the radii of these spheres for the dodecahedron and the icosahedron with edges of unit length is expressed through the *golden ratio* (Table 1.1).

Table 1.1. The golden ratio in the spheres of *dodecahedron* and *icosahedron*.

| | $R_c$ | $R_m$ | $R_i$ |
|---|---|---|---|
| Icosahedron | $\frac{1}{2}\Phi\sqrt{3-\Phi}$ | $\frac{1}{2}\Phi$ | $\frac{\frac{1}{2}\Phi^2}{\sqrt{3}}$ |
| Dodecahedron | $\frac{\Phi\sqrt{3}}{2}$ | $\frac{\Phi^2}{2}$ | $\frac{\Phi^2}{2\sqrt{3-\Phi}}$ |

Note that the ratio of the radii $\frac{R_c}{R_i}=\frac{\sqrt{3(3-\Phi)}}{\Phi}$ is the same for both icosahedron and dodecahedron. Thus, if the *dodecahedron* and *icosahedron* have the same inner spheres, their outer spheres are equal as well. This is a reflection of the hidden harmony in both *dodecahedron* and *icosahedron.*

The above examples demonstrate a crucial role of the *golden ratio* in Euclid's *Elements*, in particular in the geometry of the *Platonic solids*. This is not always understood by some modern historians of mathematics, who cannot explain for what purpose Euclid introduced the golden ratio in Book II.

## 1.4. The algebraic identities of the golden ratio

### 1.4.1. *The simplest identities*

We start with the simplest algebraic properties of the *golden ratio*. Let us represent the algebraic equation (1.9) as follows:

$$x^2 = x + 1. \tag{1.11}$$

If we substitute the root $\Phi$ (the golden ratio) instead of $x$ in equation (1.11), we then get the first remarkable property of the golden ratio:

$$\Phi^2 = \Phi + 1. \tag{1.12}$$

If we divide all terms of (1.12) by $\Phi$, we get the following expression for $\Phi$:

$$\Phi = 1 + \frac{1}{\Phi}. \tag{1.13}$$

By using (1.12) and (1.13), it is easy to prove the following identity:

$$\Phi^n = \Phi^{n-1} + \Phi^{n-2},\ n = 0, \pm1, \pm2, \pm3, \ldots. \tag{1.14}$$

### 1.4.2. *The "golden" geometric progression*

Consider the sequence of the powers of the *golden ratio*:

$$\{\ldots, \Phi^{-n}, \Phi^{-(n-1)}, \ldots, \Phi^{-2}, \Phi^{-1}, \Phi^0 = 1, \Phi^1, \Phi^2, \ldots, \Phi^{n-1}, \Phi^n, \ldots\}. \tag{1.15}$$

The sequence (1.15) has some very interesting mathematical properties. On the one hand, the sequence (1.15) is "arithmetic progression," in which each term is connected to two preceding terms by the identity (1.14); on the other hand the sequence (1.15) is "geometric progression," according to the rule:

$$\Phi^n = \Phi \times \Phi^{n-1}. \tag{1.16}$$

As every geometric progression corresponds to some logarithmic spiral, in the opinion of many researchers, the property (1.14) singles out the "*golden*" *progression* (1.15) amongst other geometric progressions. This is the cause of the widespread belief that the "*golden*" *progression* (1.15) underlies the basis of spiral forms of Nature.

### 1.4.3. *A representation of the golden ratio in "radicals"*

Let us consider once again the identity (1.12). If we take the square root on both sides of (1.12), we get the following expression:

$$\Phi = \sqrt{1 + \Phi}. \tag{1.17}$$

If we continue the substitution of (1.17) instead of $\Phi$ into the right-hand side of (1.17) ad infinitum, we then get the following astonishing representation of the *golden ratio* in the "radicals":

$$\Phi = \sqrt{1+\sqrt{1+\sqrt{1+\sqrt{1+\ldots}}}}\,. \tag{1.18}$$

### 1.4.4. *A representation of the golden ratio in the form of a continued fraction*

By using the expression (1.13), it is easy to prove the following expression of $\Phi$ in the form of a continued fraction:

$$\Phi = 1+\cfrac{1}{1+\cfrac{1}{1+\cfrac{1}{1+\cfrac{1}{1+\ldots}}}}\,. \tag{1.19}$$

The expression (1.19) has a profound mathematical sense. The surprisingly simple representation (1.19) reveals the uniqueness of the golden ratio amongst all irrational numbers from the point of view of continued fractions.

Note that the endless repetition of the same simple mathematical elements in the formulas for $\Phi$, given by (1.18) and (1.19), produce within us an "aesthetic pleasure" of rhythm and harmony.

## 1.5. Fibonacci numbers

### 1.5.1. *Fibonacci's role in the development of Western mathematics*

Leonardo Pisano Fibonacci (1175–1250) is the Italian mathematician, considered to be the most talented Western mathematician of the Middle Ages. Unfortunately, Fibonacci's contribution to the development of Western mathematics was not fully evaluated.

Figure 1.9. Leonardo Fibonacci Pisano (1170–1240).

Source: Wikipedia, The Free Encyclopedia, from http://en.wikipedia.org/wiki/Fibonacci

In his 1919 book, *Integer Number*, the Russian mathematician Professor Vasil'ev estimated the significance of Fibonacci's mathematical influence on the Western mathematics as follows:

*"The work of the learned merchant from Pisa was so far above the level of mathematical knowledge of even the scientists of his time, that its influence on mathematical literature first became noticeable two centuries after his death, at the end of the 15th century. That is when Luca Pacioli, professor at several Italian universities and Leonardo da Vinci's friend, employed many of Fibonacci's theorems and problems in his works. And then again at the beginning of the 16th century, when a group of talented Italian mathematicians, Ferro, Cardano, Tartalia, and Ferrari, formulated the beginnings of higher algebra, thanks to the solution of cubic and biquadrate equations."*

It follows from this quote that Fibonacci had surpassed the western mathematicians of that time on about two centuries. Fibonacci's historical role for western science is similar to the role of Pythagoras, who received "scientific education" from the ancient Egyptians and Babylonians and had then introduced these scientific knowledge to the Greek science. Fibonacci received his mathematical education in Arabian countries and had then introduced these in western mathematics through his mathematical works. By this he laid the foundation for further development of western mathematics.

*Fibonacci numbers* (1.1), given by the recurrent relation (1.2), are the most famous mathematical Fibonacci result, described in his book "Liber Abaci," first published in 2002.

### 1.5.2. *Kepler's formula*

Fibonacci numbers have many delightful mathematical properties, which for many centuries stimulated the mathematical imagination. Let us consider the connection of Fibonacci numbers to the golden ratio. Fractional approximation of the golden ratio (1.19) is based on Fibonacci numbers and has the following initial form:

| | |
|---|---|
| $1=\frac{1}{1}$ | (*the first approximation*) |
| $1+\frac{1}{1}=\frac{2}{1}$ | (*the second approximation*) |
| $1+\cfrac{1}{1+\cfrac{1}{1}}=\frac{3}{2}$ | (*the third approximation*) |
| $1+\cfrac{1}{1+\cfrac{1}{1}}=\frac{5}{3}$ | (*the fourth approximation*) |

By continuing this process *ad infinitum*, we find the sequence of continued fractions, which approximate the *golden proportion*; this sequence consists of the ratios of adjacent Fibonacci numbers:

$$\frac{1}{1},\frac{2}{1},\frac{3}{2},\frac{5}{3},\frac{8}{5},\frac{13}{8},\frac{21}{13},\ldots\to\Phi=\lim_{n\to\infty}\frac{F_n}{F_{n-1}}=\frac{1+\sqrt{5}}{2}. \qquad (1.20)$$

The formula (1.20) is sometimes called *Kepler's formula*. This formula expresses the famous law of *phyllotaxis* [59], according to which Nature constructs pine cones, pineapples, cacti, heads of sunflowers and other botanical objects. In other words, Nature uses the unique mathematical properties of the golden ratio (1.20) in its wonderful creations!

Albert Einstein said: "*I have deep faith that the principles of the Universe are simple and beautiful.*"

Mathematicians intuitively seek to express their mathematical results in the simplest, most compact and beautiful form. And if mathematician finds such an "aesthetic form," he gets "aesthetic pleasure." In this relation, mathematical creativity is similar to the creativity of composers or poets, whose main task is to obtain perfect musical or poetic forms, which also gives the "aesthetic pleasure."

### 1.5.3. *Variations on Fibonacci theme*

Variations on a given theme are a well-known genre in music. A distinctive feature of such musical compositions is the fact that they begin, in most cases, with one simple musical theme, which thereafter undergoes considerable changes in tempo, mood and nature. If we will follow this example of a musical piece and study a simple mathematical object, as the Fibonacci sequence (1.1), we can consider it together with its numerous variations.

For example, let us calculate *the sum of the first n Fibonacci numbers* by beginning with the simplest sums:

$$\boxed{\begin{array}{l} 1+1=2=\mathbf{3}-1 \\ 1+1+2=4=\mathbf{5}-1 \\ 1+1+2+3=7=\mathbf{8}-1 \\ 1+1+2+3+5=12=\mathbf{13}-1 \end{array}}. \quad (1.21)$$

If in these sums (1.21) we consider the numbers marked by bold: **3**, **5**, **8**, **13**,..., then it is easy to see, that they are Fibonacci numbers! We can write the sums (1.21) in the following general form:

$$F_1 + F_2 + \ldots + F_n = F_{n+2} - 1. \quad (1.22)$$

Let us consider *the sum of consecutive Fibonacci numbers with odd indices*. We start with the simplest sums:

$$\boxed{\begin{array}{l} 1+2=\mathbf{3} \\ 1+2+5=\mathbf{8} \\ 1+2+5+13=\mathbf{21} \\ 1+2+5+13+34=\mathbf{55} \end{array}}. \quad (1.23)$$

The following general formula is derived from (1.23):

$$F_1 + F_3 + F_5 + \ldots + F_{2n-1} = F_{2n}. \tag{1.24}$$

*The sum of the consecutive Fibonacci numbers with even indices* is the following:

$$F_2 + F_4 + F_6 + \ldots + F_{2n} = F_{2n+1} - 1. \tag{1.25}$$

Let us calculate *the sum of the squares of the n consecutive Fibonacci numbers*:

$$F_1^2 + F_2^2 + \ldots + F_n^2. \tag{1.26}$$

We start from the analysis of the simplest sums of the kind (1.26):

$$\boxed{\begin{array}{l} 1^2+1^2=2=\mathbf{1}\times\mathbf{2} \\ 1^2+1^2+2^2=6=\mathbf{2}\times\mathbf{3} \\ 1^2+1^2+2^2+3^2=15=\mathbf{3}\times\mathbf{5} \\ 1^2+1^2+2^2+3^2+5^2=40=\mathbf{5}\times\mathbf{8} \end{array}}. \tag{1.27}$$

The analysis of (1.27) leads us to the following general formula:

$$F_1^2 + F_2^2 + \ldots + F_n^2 = F_n F_{n+1}. \tag{1.28}$$

Let us consider *the sum of the squares of two adjacent Fibonacci numbers*:

$$F_{n-1}^2 + F_n^2. \tag{1.29}$$

We start from the analysis of the simplest sums of the kind (1.29):

$$\boxed{\begin{array}{l} 1^2+1^2=1+1=\mathbf{2} \\ 1^2+2^2=1+4=\mathbf{5} \\ 2^2+3^2=4+9=\mathbf{13} \\ 3^2+5^2=9+25=\mathbf{34} \end{array}}. \tag{1.30}$$

The analysis of (1.30) leads us to the following general formula:

$$F_{n-1}^2 + F_n^2 = F_{2n-1}. \tag{1.31}$$

### 1.5.4. *The "extended" Fibonacci numbers*

Fibonacci numbers (1.1) can be extended for the negative values of the indices of $n$ (see Table 1.2).

Table 1.2. The "extended" Fibonacci numbers.

| $n$ | 0 | 1 | 2 | 3 | 4 | 5 | 6 | 7 | 8 | 9 | 10 |
|---|---|---|---|---|---|---|---|---|---|---|---|
| $F_n$ | 0 | 1 | 1 | 2 | 3 | 5 | 8 | 13 | 21 | 34 | 55 |
| $F_{-n}$ | 0 | 1 | –1 | 2 | –3 | 5 | –8 | 13 | –21 | 34 | -55 |

The "extended" Fibonacci numbers are connected by the following simple relationship:

$$F_{-n} = (-1)^{n+1} F_n. \tag{1.32}$$

### 1.5.5. *Cassini's formula for Fibonacci numbers*

The history of science does not reveal why French astronomer Giovanni Domenico Cassini (1625–1712) had been interested in Fibonacci numbers. Most likely it was simply an object of rapture for the great astronomer. At that time many serious scientists were fascinated by Fibonacci numbers and the golden ratio. We recall that Fibonacci numbers and the golden ratio were an aesthetic object for Johannes Kepler, who was a contemporary of Cassini.

Consider now the Fibonacci sequence: 1,1,2,3,5,8,13,21,34,.... Take the Fibonacci number 5 and square it, that is, $5^2 = 25$. Next consider the product of the two Fibonacci numbers 3 and 8, which are adjacent to 5, that is, $3 \times 8 = 24$. Then we can write:

$$5^2 - 3 \times 8 = 1.$$

Note that the difference obtained is equal to (+1).

Now we make the same procedure for the next Fibonacci number 8, that is, we first square it ($8^2 = 64$), then we calculate the product of the two Fibonacci numbers 5 and 13, which are adjacent to 8, that is, $5 \times 13 = 65$. After a comparison of the product $5 \times 13 = 65$ to the square $8^2 = 64$ we get:

$$8^2 - 5 \times 13 = -1.$$

Note that the obtained difference is equal to (−1).

Furthermore we have:

$$13^2 - 8 \times 21 = 1;$$

$$21^2 - 13 \times 34 = -1$$

and so on.

We note that the square of any Fibonacci number $F_n$ always differs from the product of the its two adjacent Fibonacci numbers $F_{n-1}$ and $F_{n+1}$ by 1. This difference is alternately +1 or −1 up to infinity and depends on the index $n$ of the initial Fibonacci number $F_n$. If the index $n$ is even, then the result is −1, and if odd, +1. This property of Fibonacci numbers can be expressed by the following mathematical formula:

$$F_n^2 - F_{n-1}F_{n+1} = (-1)^{n+1}. \quad (1.33)$$

This wonderful formula evokes spiritual thrill when we realize that the identity (1.33) is valid for any value of integer $n$ from $-\infty$ up to $+\infty$. Thus this endless alternation of +1 and −1 in the expression (1.33) evokes a genuine aesthetic feeling of rhythm and harmony.

### 1.5.6. *Fibonacci λ-numbers*

Let us give a positive integer $\lambda = 1, 2, 3, \ldots$ and consider the following recurrent relation [32]:

$$F_\lambda(n+2) = \lambda F_\lambda(n+1) + F_\lambda(n); \quad F_\lambda(0) = 0, F_\lambda(1) = 1. \quad (1.34)$$

The recurrent relation (1.34) "generates" an infinite number of new numerical sequences, because every positive integer $\lambda = 1, 2, 3, \ldots$ "generates" its own numerical sequence.

Let us consider the partial cases of the recurrent relation (1.34). For the case $\lambda = 1$, the recurrent relation (1.34) is reduced to the following recurrent relation:

$$F_1(n+2) = F_1(n+1) + F_1(n); \quad F_1(0) = 0, F_1(1) = 1. \quad (1.35)$$

This recurrent relation "generates" the classic Fibonacci numbers: 0,1,1,2,3,5,8,13,21,34,….

Based on this fact, we will name the numerical sequences, generated by the recurrence relation (1.34), the *Fibonacci λ-numbers* [32].

For $\lambda = 2$ the recurrent relation (1.34) is reduced to the recurrent relation:

$$F_2(n+2) = 2F_2(n+1) + F_2(n); \quad F_2(0) = 0, F_2(1) = 1, \tag{1.36}$$

which gives the so-called *Pell numbers* [60]:

$$0,1,2,5,12,29,70,\ldots. \tag{1.37}$$

For $\lambda = 3$ the recurrent relation (1.34) is reduced to the following recurrent relation:

$$F_3(n+2) = 3F_3(n+1) + F_3(n); \quad F_3(0) = 0, F_3(1) = 1. \tag{1.38}$$

The Fibonacci λ-numbers have many remarkable properties, similar to the properties of the classic Fibonacci numbers. It is proved in [32] that the Fibonacci λ-numbers, similarly to the classic Fibonacci numbers can be "extended" to the negative values of the indices $n$.

Table 1.3 shows the four "extended" sequences of the Fibonacci λ-numbers, corresponding to the values $\lambda = 1, 2, 3, 4$.

Table 1.3. The "extended" Fibonacci λ-numbers ($\lambda = 1, 2, 3, 4$).

| $n$ | 0 | 1 | 2 | 3 | 4 | 5 | 6 | 7 | 8 |
|---|---|---|---|---|---|---|---|---|---|
| $F_1(n)$ | 0 | 1 | 1 | 2 | 3 | 5 | 8 | 13 | 21 |
| $F_1(-n)$ | 0 | 1 | –1 | 2 | –3 | 5 | –8 | 13 | –21 |
| $F_2(n)$ | 0 | 1 | 2 | 5 | 12 | 29 | 70 | 169 | 408 |
| $F_2(-n)$ | 0 | 1 | –2 | 5 | –12 | 29 | –70 | 169 | –408 |
| $F_3(n)$ | 0 | 1 | 3 | 10 | 33 | 109 | 360 | 1189 | 3927 |
| $F_3(-n)$ | 0 | 1 | –3 | 10 | –33 | 109 | –360 | 1199 | –3927 |
| $F_4(n)$ | 0 | 1 | 4 | 17 | 72 | 305 | 1292 | 5473 | 23184 |
| $F_4(-n)$ | 0 | 1 | –4 | 17 | –72 | 305 | –1292 | 5473 | –23184 |

### 1.5.7. *The generalized Cassini's formula for the Fibonacci λ-numbers*

It is easy to prove [32] the following *generalized Cassini's formula* for the Fibonacci λ-numbers:

$$F_\lambda^2(n) - F_\lambda(n-1)F_\lambda(n+1) = (-1)^{n+1}. \tag{1.39}$$

Consider the examples of the validity of the identity (1.39) for the various sequences shown in Table 1.3. Let us consider the $F_2(n)$-sequence for the case $n = 7$. For this case we should consider the following triple of the Fibonacci 2-numbers $F_2(n)$: $F_2(6) = 70$, $F_2(7) = 169$, $F_2(8) = 408$. By performing calculations over them according to (1.39), we get the following result:

$$(169)^2 - 70 \times 408 = 28561 - 28560 = 1,$$

which corresponds to the identity (1.39), because for the case $n = 7$ we have: $(-1)^{n+1} = (-1)^8 = 1$.

Now let us consider the $F_3(n)$-sequence from Table 1.3 for $n = 6$. We should choose the following triple of the Fibonacci 3-numbers $F_3(n)$:

$$F_3(5) = 109,\ F_3(6) = 360,\ F_3(7) = 1189.$$

By performing calculations over them according to (1.39), we get the following result:

$$(360)^2 - 109 \times 1189 = 129600 - 129601 = -1,$$

which corresponds to the identity (1.39), because for $n = 6$ we have $(-1)^{n+1} = (-1)^7 = -1$.

Finally, let us consider the $F_4(-n)$-sequence from Table 1.3 for $n = -5$. For this case we should choose the following triple of the Fibonacci 4-numbers $F_4(-n)$:

$$F_4(-4) = -72,\ F_4(-5) = 305,\ F_4(-6) = -1292.$$

By performing calculations over them according to (1.39), we get the following result:

$$(305)^2 - (-72) \times (-1292) = 93025 - 93024 = 1,$$

which corresponds to the identity (1.39), because for $n = -5$ we have $(-1)^{n+1} = (-1)^{-4} = 1$.

Thus, by studying the generalized Cassini formula (1.39) for the Fibonacci λ-numbers, we came to the discovery of an infinite number of integer recurrent sequences in the range from $+\infty$ to $-\infty$ with the following unique mathematical property, expressed by the generalized Cassini formula (1.39), which sounds as follows:

*The quadrate of any* Fibonacci λ-number $F_\lambda(n)$ *is always different from the product of the two adjacent Fibonacci λ-numbers* $F_\lambda(n-1)$ *and* $F_\lambda(n+1)$, *which surround* the initial Fibonacci λ-number $F_\lambda(n)$, *by the number* 1; herewith *the sign of the difference of* 1 *depends on the parity of n: if n is even,* then the difference of 1 *is taken with the sign* "minus", *otherwise*, with the sign "*plus*."

Until now, we have assumed that only the classic Fibonacci numbers are of the unusual property, given by the Cassini formula (1.33). However, as it is shown above, the number of such numerical sequences is infinite. All the Fibonacci λ-numbers, given by (1.34), are of a similar property, given by the generalized Cassini formula (1.39)!

### 1.5.8. *Fibonacci Q-matrix*

In recent decades the theory of Fibonacci numbers has been supplemented by the theory of the so-called *Fibonacci Q-matrix* [54]. This is the (2 × 2)-matrix of the following form:

$$Q = \begin{pmatrix} 1 & 1 \\ 1 & 0 \end{pmatrix}. \tag{1.40}$$

The Fibonacci $Q$-matrix has the following remarkable properties [54]:

$$Q^n = \begin{pmatrix} F_{n+1} & F_n \\ F_n & F_{n-1} \end{pmatrix},$$

$$\det Q = -1, \;\; \det Q^n = F_{n-1}F_{n+1} - F_n^2 = (-1)^n,$$

| $n$ | 0 | 1 | 2 | 3 | 4 | 5 |
|---|---|---|---|---|---|---|
| $Q^n$ | $\begin{pmatrix}1 & 0\\0 & 1\end{pmatrix}$ | $\begin{pmatrix}1 & 1\\1 & 0\end{pmatrix}$ | $\begin{pmatrix}2 & 1\\1 & 1\end{pmatrix}$ | $\begin{pmatrix}3 & 2\\2 & 1\end{pmatrix}$ | $\begin{pmatrix}5 & 3\\3 & 2\end{pmatrix}$ | $\begin{pmatrix}8 & 5\\5 & 3\end{pmatrix}$ |
| $Q^{-n}$ | $\begin{pmatrix}1 & 0\\0 & 1\end{pmatrix}$ | $\begin{pmatrix}0 & 1\\1 & -1\end{pmatrix}$ | $\begin{pmatrix}1 & -1\\-1 & 2\end{pmatrix}$ | $\begin{pmatrix}-1 & 2\\2 & -3\end{pmatrix}$ | $\begin{pmatrix}2 & -3\\-3 & 5\end{pmatrix}$ | $\begin{pmatrix}-3 & 5\\5 & -8\end{pmatrix}$ |

## 1.6. Lucas numbers

### 1.6.1. *François Édouard Anatole Lucas*

Fibonacci didn't study the mathematical properties of the numerical sequence (1.1), for which he became famous. This was done later by other mathematicians. Since the 19th century the mathematical works, devoted to Fibonacci numbers, "began to reproduce like Fibonacci's rabbits" according to the witty expression of one mathematician. The French mathematician François Édouard Anatole Lucas was born in 1842 and became the leader of 19th century in Fibonacci studies. Unfortunately he passed away prematurely in 1891 as the result of a freak accident at a banquet when a dish was smashed and a splinter wounded his cheek. He died from an infection a few days later.

Figure 1.10. François Édouard Anatole Lucas (1842–1891).
Source: Wikipedia, The Free Encyclopedia, fromhttps://en.wikipedia.org/wiki/Felix_Klein

### 1.6.2. *The Lucas numbers*

Lucas introduced the concept of generalized Fibonacci numbers, which can be calculated by the following general recurrent relation:

$$G_n = G_{n-1} + G_{n-2} \tag{1.41}$$

for arbitrary initial terms (seeds) $G_1$ and $G_2$.

However, Lucas had introduced the so-called Lucas numbers (1.4), which are generated by the recurrent relation (1.3).

If we analyze the Lucas numbers (1.4), generated by the recurrent relation (1.3), similarly to the Fibonacci numbers (1.1), the following identities for Lucas numbers emerge:

$$L_1 + L_3 + L_5 + \ldots + L_{2n-1} = L_{2n} - 2$$

$$L_2 + L_4 + L_6 + \ldots + L_{2n} = L_{2n+1} - 1$$

$$L_1^2 + L_2^2 + \ldots + L_n^2 = L_n L_{n+1} - 2$$

$$L_n^2 + L_{n+1}^2 = 5F_{2n+1}$$

$$\lim_{n\to\infty} \frac{L_n}{L_{n-1}} = \Phi = \frac{1+\sqrt{5}}{2}.$$

### 1.6.3. *The "extended" Lucas numbers*

Similarly to the Fibonacci numbers (Table 1.2), the Lucas numbers (1.4) can be extended for the negative values of the indices $n$ (Table 1.4).

Table 1.4. The "extended" Lucas numbers.

| $n$ | 0 | 1 | 2 | 3 | 4 | 5 | 6 | 7 | 8 | 9 | 10 |
|---|---|---|---|---|---|---|---|---|---|---|---|
| $L_n$ | 2 | 1 | 3 | 4 | 7 | 11 | 18 | 29 | 47 | 76 | 123 |
| $L_{-n}$ | 2 | –1 | 3 | –4 | 7 | –11 | 18 | –29 | 47 | –76 | 123 |

The "extended" Lucas numbers are connected by the following simple relation:

$$L_{-n} = (-1)^n L_n. \tag{1.42}$$

## 1.7. Binet's formulas

### 1.7.1. *Jacques Philippe Marie Binet*

Besides Lucas, the French 19th century mathematician Jacques Philippe Marie Binet (1786–1856) was another enthusiast of Fibonacci numbers and the golden ratio.

Figure 1.11. Jacques Philippe Marie Binet (1786–1856).
Source: Wikipedia, The Free Encyclopedia, from
http://en.wikipedia.org/wiki/Jacques_Philippe_Marie_Binet

### 1.7.2. *Binet's formulas for Fibonacci and Lucas numbers*

In the theory of Fibonacci and Lucas numbers [53–55], the following formula is well-known:

$$\Phi^n = \frac{L_n + F_n\sqrt{5}}{2}. \tag{1.43}$$

It links the power of the golden ratio $\Phi^n$ ($n$) = 0, 1, 2, 3,...) with the "extended" Fibonacci and Lucas numbers $F_n$ and $L_n$.

By using formula (1.43), we can represent the "extended" Fibonacci and Lucas numbers through the *golden ratio*. For this purpose it is enough to write the formulas for the sum and the difference of the $n$th degrees of the golden ratio $\Phi^n + \Phi^{-n}$ and $\Phi^n - \Phi^{-n}$ as follows:

$$\Phi^n + \Phi^{-n} = \frac{(L_n + L_{-n}) + (F_n + F_{-n})\sqrt{5}}{2}, \tag{1.44}$$

$$\Phi^n - \Phi^{-n} = \frac{(L_n - L_{-n}) + (F_n - F_{-n})\sqrt{5}}{2}. \tag{1.45}$$

Let us consider (1.44) and (1.45) for the even values of the indices $n = 2k$. From Tables 1.2 and 1.4 we have the following simple forms, respectively:

$$\Phi^{2k} + \Phi^{-2k} = L_{2k}, \tag{1.46}$$

$$\Phi^{2k} - \Phi^{-2k} = F_{2k}\sqrt{5}. \tag{1.47}$$

For the odd indices $n = 2k + 1$ we can represent the formulas (1.44) and (1.45) as follows, respectively:

$$\Phi^{2k+1} + \Phi^{-(2k+1)} = F_{2k+1}\sqrt{5}, \tag{1.48}$$

$$\Phi^{2k+1} - \Phi^{-(2k+1)} = L_{2k+1}. \tag{1.49}$$

By uniting formulas (1.47), (1.48) and formulas (1.46), (1.49), we can now represent the "extended" Fibonacci and Lucas numbers in the following compact forms [15]:

$$F_n = \begin{cases} \dfrac{\Phi^n + \Phi^{-n}}{\sqrt{5}} \text{ for } n = 2k+1 \\ \dfrac{\Phi^n - \Phi^{-n}}{\sqrt{5}} \text{ for } n = 2k, \end{cases} \tag{1.50}$$

$$L_n = \begin{cases} \Phi^n + \Phi^{-n} \text{ for } n = 2k \\ \Phi^n - \Phi^{-n} \text{ for } n = 2k+1. \end{cases} \tag{1.51}$$

The formulas (1.50) and (1.51) evoke a feeling of "aesthetic pleasure," once again convincing us in the sublime power of mathematics! We know that the "extended" Fibonacci and Lucas numbers are always integers, but any degree of the golden ratio is an irrational number. It follows from this that the integer numbers $F_n$ and $L_n$ can be represented by using formulas (1.50) and (1.51) through the golden ratio $\Phi$!

## 1.8. Pascal's triangle and Fibonacci *p*-numbers

### 1.8.1. *Mathematical discovery by George Polya*

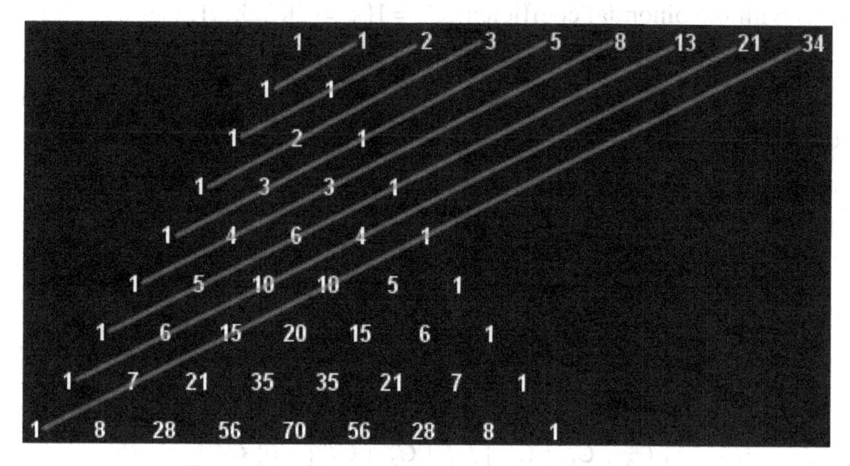


Figure 1.12. Fibonacci numbers in Pascal's triangle.
Source: Pascal's Triangle http://www.goldennumber.net/pascals-triangle/

As it is known, Pascal triangle plays an important role in the combinatorial analysis and has many interesting applications in mathematics and computer science, in particular, in coding theory.

By studying the so-called diagonal sums of Pascal's triangle, American mathematician George Polya came to a very simple and unexpected discovery, described in the book [61] (see Fig. 1.12).

It should be noted that this very simple mathematical result during many centuries was not known to Blaise Pascal and other mathematicians, who studied Fibonacci numbers and combinatorial analysis.

### 1.8.2. *The rectangular Pascal triangle*

There are many various forms of Pascal triangle representation, for example, in the form of the isosceles triangle (Fig. 1.12), in the form of rectangular table ("Tartalja's Rectangle"), etc. We will consider the so-called *rectangular Pascal triangle* [14] that can be represented as the following table of binomial coefficients (Table 1.5).

The rows of the rectangular Pascal triangle (Table 1.5) are numbered from above to the bottom. The binomial coefficients $C_0^0 = C_1^0 = C_2^0 = ... = C_n^0 = 1$ are forming the "zero" row. Every $k$th row starts with the binomial coefficient $C_k^k = 1(k = 0,1,2,3,...)$.

The columns of Pascal triangle are numbered from left to right; the last left-hand column, which consists of the only binomial coefficient ($C_0^0$ = 1) is called *zero-column*. The $n$th column ($n$ = 0, 1, 2, 3, ...) includes the following binomial coefficients:

$$C_n^0, C_n^1, C_n^2, ..., C_n^k, ..., C_n^{n-k}, ..., C_n^n,$$

where $C_n^k = C_n^{n-k}$.

Table 1.5. Rectangular Pascal triangle.

| $C_0^0$ | $C_1^0$ | $C_2^0$ | $C_3^0$ | $C_4^0$ | $C_5^0$ | ... | $C_n^0$ |
|---|---|---|---|---|---|---|---|
| | $C_1^1$ | $C_2^1$ | $C_3^1$ | $C_4^1$ | $C_5^1$ | ... | $C_n^1$ |
| | | $C_2^2$ | $C_3^2$ | $C_4^2$ | $C_5^2$ | ... | $C_n^2$ |
| | | | $C_3^3$ | $C_4^3$ | $C_5^3$ | ... | $C_n^3$ |
| | | | | $C_4^4$ | $C_5^4$ | ... | $C_n^4$ |
| | | | | | $C_5^5$ | ... | $C_n^5$ |
| | | | | | | | ⋮ |
| | | | | | | | $C_n^n$ |

The binomial coefficients and Pascal triangle are widely used in various fields of mathematics and computer science. The well-known mathematician Jacob Bernoulli (1655–1705) wrote:

*"This table has a number of wonderful properties. Just now we have shown that it expresses an essence of connections theory, but those scientists, who closely adjoin to geometry, know that this table preserves a number of fundamental secrets of this area of mathematics."*

Figure 1.12 shows fundamental connection between Pascal triangle and Fibonacci numbers, described in Polya's book [61]. A generalization of this regularity led to the new recurrent sequences, named *generalized Fibonacci numbers* or *Fibonacci p-numbers* [14].

### 1.8.3. *Definition of Fibonacci p-numbers*

Consider now the rectangular Pascal triangle (Table 1.5), represented in the numerical form (Table 1.6).

We can name the given table of binomial coefficients as *Pascal 0-triangle* (the meaning of such definition will become clear below). If we summarize now the binomial coefficients of Pascal 0-triangle by columns starting from the 0-column, then we get the "binary sequence":

$$1, 2, 4, 8, 16, 32, 64, \ldots, 2^{n-1}, \ldots \quad (1.52)$$

Table 1.6. Pascal 0-triangle.

| | | | | | | | | | |
|---|---|---|---|---|---|---|---|---|---|
| 1 | 1 | 1 | 1 | 1 | 1 | 1 | 1 | 1 | 1 |
| | 1 | 2 | 3 | 4 | 5 | 6 | 7 | 8 | 9 |
| | | 1 | 3 | 6 | 10 | 15 | 21 | 28 | 36 |
| | | | 1 | 4 | 10 | 20 | 35 | 56 | 84 |
| | | | | 1 | 5 | 15 | 35 | 70 | 126 |
| | | | | | 1 | 6 | 21 | 56 | 126 |
| | | | | | | 1 | 7 | 28 | 84 |
| | | | | | | | 1 | 8 | 36 |
| | | | | | | | | 1 | 9 |
| | | | | | | | | | 1 |
| 1 | 2 | 4 | 8 | 16 | 32 | 64 | 128 | 256 | 512 |

And now we perform some "manipulations" over the Pascal 0-triangle. Let us shift each row of the Pascal 0-triangle (Table 1.6) on one column to the right relatively to the previous row. As a result of such transformations we get Table 1.7 called Pascal 1-triangle.

Table 1.7. Pascal 1-triangle.

| 1 | 1 | 1 | 1 | 1 | 1 | 1 | 1 | 1 | 1 | 1 | 1 |
|---|---|---|---|---|---|---|---|---|---|---|---|
| | | 1 | 2 | 3 | 4 | 5 | 6 | 7 | 8 | 9 | 10 |
| | | | | 1 | 3 | 6 | 10 | 15 | 21 | 28 | 36 |
| | | | | | | 1 | 4 | 10 | 20 | 35 | 56 |
| | | | | | | | | 1 | 5 | 15 | 35 |
| | | | | | | | | | | 1 | 6 |
| 1 | 1 | 2 | 3 | 5 | 8 | 13 | 21 | 34 | 55 | 89 | 144 |

And now we summarize the binomial coefficients of the Pascal 1-triangle by columns. To our amazement, we find that such summation results in Fibonacci numbers:

$$1, 1, 2, 3, 5, 8, 13, \ldots, F_n, \ldots, \tag{1.53}$$

where $F_n$ is the $n$th Fibonacci number, given by the recurrent relation (1.2).

If we shift the binomial coefficients of each row in $p$ columns to the right relatively to the previous row ($p = 0, 1, 2, 3, \ldots$) in Table 1.6, we get the numerical table named *Pascal p-triangle*.

It is clear that Pascal 0-triangle, that is, Pascal $p$-triangle, corresponding to $p = 0$, is the initial Pascal triangle (Table 1.6). The Pascal 1-triangle is represented in Table 1.7. The Pascal 2-triangle is represented in Table 1.8.

Table 1.8. Pascal 2-triangle.

| 1 | 1 | 1 | 1 | 1 | 1 | 1 | 1 | 1 | 1 | 1 | 1 | 1 |
|---|---|---|---|---|---|---|---|---|---|---|---|---|
| | | | 1 | 2 | 3 | 4 | 5 | 6 | 7 | 8 | 9 | 10 |
| | | | | | | 1 | 3 | 6 | 10 | 15 | 21 | 28 |
| | | | | | | | | | 1 | 4 | 10 | 20 |
| | | | | | | | | | | | | 1 |
| 1 | 1 | 1 | 2 | 3 | 4 | 6 | 9 | 13 | 19 | 28 | 41 | 60 |

And now we summarize by columns the binomial coefficients of Pascal 2-triangle. As an outcome, we get the following numerical sequence:

$$1, 1, 1, 2, 3, 4, 6, 9, 13, 19, \ldots, F_2(n), \ldots. \tag{1.54}$$

Denote by $F_2(n)$ the $n$th number of the sequence (1.54), starting from the left. It is easy to see the following regularity in the numerical sequence (1.54), which is expressed by the recurrent relation

$$F_2(n) = F_2(n-1) + F_2(n-3) \text{ for } n \geq 4 \tag{1.55}$$

at the seeds

$$F_2(1) = F_2(2) = F_2(3) = 1. \tag{1.56}$$

The sequence (1.54) is called *Fibonacci 2-numbers* [14].

In general case (for the given $p = 0, 1, 2, 3, \ldots$), after the summation of the binomial coefficients of the Pascal $p$-triangle by columns, we get the numerical sequence, given by the following recurrent relation:

$$F_p(n) = F_p(n-1) + F_p(n-p-1) \quad \text{for } n > p+1 \tag{1.57}$$

for the seeds

$$F_p(1) = F_p(2) = \ldots = F_p(p+1) = 1. \tag{1.58}$$

The numerical sequences, generated by the recurrent relation (1.57) at the seeds (1.58) are named in [14] the *Fibonacci p-numbers*.

It is proved above that for the case $p = 0$ the *Fibonacci p-numbers* are reduced to the binary sequence (1.52) and for the case $p = 1$ to the classic Fibonacci numbers (1.53). For the case $p = \infty$, the *Fibonacci p-numbers* are reduced to the following trivial sequence:

$$\{1, 1, 1, \ldots, 1, \ldots\}. \tag{1.59}$$

### 1.8.4. *A representation of the Fibonacci p-numbers through binomial coefficients*

By analyzing Pascal 0-triangle (Table 1.5), it is easy to derive the mathematical formula, representing the Fibonacci 0-numbers (the binary numbers (1.52)) by the binomial coefficients:

$$2^n = C_n^0 + C_n^1 + ... + C_n^n. \tag{1.60}$$

The formula (1.60) is the basis of the binary code, which underlies modern computers. Really, let us consider the set of the binary $n$-digit words, starting from the code combination 00...0 up to the code combination 11...1. As is known, the number of elements of this set is equal to $2^n$. Let us divide this set into $(n + 1)$ disjoint subsets, starting from the first subset, which consists only from the bits of 0. It is clear that the only code word 00...0 belongs to this subset, that is, the number of elements of this subset is equal to $C_n^0 = 1$. Then, we can consider the second subset from the binary words, which contain only 1 and the $(n - 1)$ 0's. It is clear that the number of the elements of this subset is equal to $C_n^1$. We can consider the $(m + 1)$th subset from the $n$-digit binary words, which consist of the $m$ 1's and $(n - m)$ 0's; the number of the code words of this subset is equal to $C_n^m$. At last, we consider the $(n + 1)$th subset from the binary word, which consists only from 1's. It is clear that the only code word 11...1 satisfies this condition, that is, the number of elements of this subset is equal to $C_n^n = 1$. It follows from this reasoning a correctness of formula (1.60).

Studying the Pascal $p$-triangle [14], we can represent the generalized Fibonacci $p$-number $F_n(n + 1)$, given by the recurrent relation (1.57) at the seeds (1.58), through the binomial coefficients as follows:

$$F_p(n+1) = C_n^0 + C_{n-p}^1 + C_{n-2p}^2 + C_{n-3p}^3 + C_{n-4p}^4 + \dots. \tag{1.61}$$

Note that the known formula (1.60) is a partial case of (1.61) for the case $p = 0$. For the cases $p = 1$, formula (1.61) is reduced to the following formula, which connects the classic Fibonacci numbers $F_{n+1} = F_1(n + 1)$ to binomial coefficients:

$$F_{n+1} = F_1(n+1) = C_n^0 + C_{n-1}^1 + C_{n-2}^2 + C_{n-3}^3 + C_{n-4}^4 + \dots. \tag{1.62}$$

It is clear that formulas (1.61), (1.62) are another confirmation of the deep connection between the theory of Fibonacci numbers and combinatorial analysis.

## 1.9. The golden *p*-proportions

### 1.9.1. *A ratio of the adjacent Fibonacci p-numbers*

The so-called *Kepler's formula*, which gives relationships between the golden ratio and Fibonacci numbers, is well-known:

$$\Phi = \underbrace{\lim}_{n\to\infty} \frac{F_n}{F_{n-1}} = \frac{1+\sqrt{5}}{2}. \tag{1.63}$$

A question arises: what is the limit of the ratio of two adjacent Fibonacci *p*-numbers? Now we introduce the following definition:

$$\underbrace{\lim}_{n\to\infty} \frac{F_p(n)}{F_p(n-1)} = x. \tag{1.64}$$

By using the recurrent relation (1.57), we can represent the ratio of the two adjacent Fibonacci *p*-numbers as follows:

$$\frac{F_p(n)}{F_p(n-1)} = \frac{F_p(n-1)+F_p(n-p-1)}{F_p(n-p-1)}$$
$$= 1 + \frac{1}{\frac{F_p(n-1)}{F_p(n-p-1)}} = 1 + \frac{1}{\frac{F_p(n-1)\cdot F_p(n-2)\cdots F_p(n-p)}{F_p(n-2)\cdot F_p(n-3)\ \cdots\ F_p(n-p-1)}}. \tag{1.65}$$

Taking into consideration (1.65), for the case $n\to\infty$, we can replace the expression (1.65) by the following algebraic equation:

$$x^{p+1} = x^p + 1. \tag{1.66}$$

Denote by $\Phi_p$ ($p = 0, 1, 2, 3,\ldots$) a positive root of the algebraic equation (1.66). Let us consider now the partial cases of Eq. (1.66) for the different values of $p$. For $p = 0$, Eq. (1.66) is reduced to the trivial equation: $x = 2$. For $p = 1$, Eq. (1.66) is reduced to the algebraic equation of the golden ratio:

$$x^2 = x + 1 \tag{1.67}$$

with positive root $\Phi = \frac{1+\sqrt{5}}{2}$ (the golden ratio).

### 1.9.2. *A generalization of Euclid's division in extreme and mean ratio (DEMR)*

Equation (1.66) has the following geometric interpretation. Let us give the integer non-negative number $p$ ($p$ = 0, 1, 2, 3, ...) and divide a line $AB$ by a point $C$ in the following proportion (Fig. 1.13):

$$\frac{CB}{AC}=\left(\frac{AB}{CB}\right)^{p} \tag{1.68}$$

where $p$ = 0, 1, 2, 3, ... .

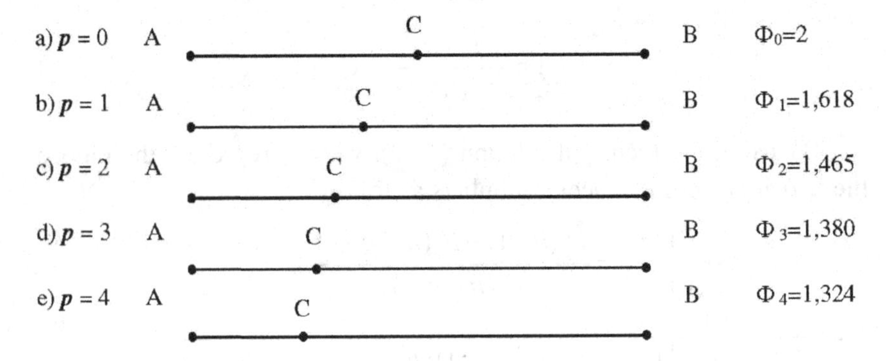


Figure 1.13. The golden $p$-sections.

Note that the proportion (1.68) is reduced to the "dichotomy" for the case $p$ = 0 (Fig. 1.13(a)) and to the classic "division in the extreme and mean ratio" (the golden ratio) for the case $p$ = 1 (Fig. 1.13(b)). Taking into consideration this fact, we will name the division of the line segment $AB$ by the point $C$ in the proportion (1.68) the *golden p-section* and the positive root of Eq. (1.66) the *golden p-proportion* [14].

### 1.9.3. *The simplest algebraic properties of the golden p-proportions*

If we substitute the golden $p$-proportion $\Phi_n$ instead $x$ in Eq. (1.66), we get the following identity for the golden $p$-proportion:

$$\Phi_p^{p+1}=\Phi_p^p+1. \tag{1.69}$$

If we divide all terms of the identity (1.69) by $\Phi_p^p$, we get the following identities for the golden $p$-proportion:

$$\Phi_p = 1 + \frac{1}{\Phi_p^p} \tag{1.70}$$

or

$$\Phi_p - 1 = \frac{1}{\Phi_p^p}. \tag{1.71}$$

Note that for the case $p = 0$ ($\Phi_0 = 2$) the identities (1.70) and (1.71) are reduced to the following trivial expressions:

$$2 = 1 + \frac{1}{1} \text{ and } 2 - 1 = \frac{1}{1}.$$

For the case $p = 1$, we have: $\Phi_1 = \Phi = \frac{1+\sqrt{5}}{2}$ and the identities (1.70) and (1.71) are reduced to the well-known identities (1.12) $\Phi^2 = \Phi + 1$ and (1.13) $\Phi = 1 + \frac{1}{\Phi}$.

If we multiply and divide repeatedly all terms of the identity (1.70) by $\Phi_p$, we get the following remarkable identities, connecting the powers of the golden $p$-proportion:

$$\Phi_p^n = \Phi_p^{n-1} + \Phi_p^{n-p-1} = \Phi_p \times \Phi_p^{n-1} \left(n = 0, \pm1, \pm2, \pm3, \ldots\right). \tag{1.72}$$

Note that for the case $p = 0$, $\Phi_p = \Phi_0 = 2$ and then the identities (1.72) are reduced to the following trivial identities for the "binary" numbers:

$$2^n = 2^{n-1} + 2^{n-1} = 2 \times 2^{n-1}.$$

For the case $p = 1$ we have: $\Phi_1 = \Phi = \frac{1+\sqrt{5}}{2}$ and then the identities (1.72) are reduced to the following well-known identities for the classic golden ratio:

$$\Phi^n = \Phi^{n-1} + \Phi^{n-2} = \Phi \times \Phi^{n-1}. \tag{1.73}$$

## 1.10. Phyllotaxis as the main reason for the creation of Fibonacci number theory in modern mathematics

### 1.10.1. *What is phyllotaxis*?

Among the natural phenomena, which surround us, perhaps, the *botanical phenomenon of phyllotaxis* is the best known and most common. This phenomenon is inherent to many biological objects. The essence of phyllotaxis consists in a spiral disposition of leaves on plant's stems of trees, petals in flower baskets, seeds in pine cones and sunflower heads etc. This phenomenon, known already since Kepler's time, was a subject of discussion for many scientists, including Leonardo da Vinci, Turing, Veil and so on. In phyllotaxis phenomenon more complex concepts of symmetry are used, in particular, a concept of *helical symmetry*.

The phenomenon of phyllotaxis reveals itself especially in florescences and densely packed botanical structures such as pine cones, pineapples, cacti, heads of sunflower and cauliflower and many other botanical objects (Fig. 1.14).

Figure 1.14. Phyllotaxis structures: (a) cactus; (b) head of sunflower; (c) coneflower; (d) Romanescue cauhflower; (e) pineapple; (f) pinecone.

On the surfaces of such objects, their bio-organs (seeds on the disks of sunflower heads and pine cones etc.) laid down in left-twisted and right-twisted spirals. For such phyllotaxis objects, most recognized are the numerical ratios of the left-handed and right-handed spirals, observed on the surface of the phyllotaxis objects. Botanists have proved that these ratios are equal to the ratios of the adjacent Fibonacci numbers, that is,

$$\frac{F_{n+1}}{F_n}: \quad \frac{2}{1}, \frac{3}{2}, \frac{5}{3}, \frac{8}{5}, \frac{13}{8}, \frac{21}{13}, \ldots \to \Phi = \frac{1+\sqrt{5}}{2}. \tag{1.74}$$

The ratios (1.74) are called *phyllotaxis orders* [59]. They are different for different phyllotaxis objects. For example, heads of sunflower can have the phyllotaxis orders given by Fibonacci's ratios $\frac{89}{55}, \frac{144}{89}$ and even $\frac{233}{144}$.

### 1.10.2. *Fibonacci Association*

The studies of French mathematicians of 19th century Lucas and Binet stimulated further research of Fibonacci numbers in modern mathematics. In 1963 a group of U.S. mathematicians created the Fibonacci Association. In the same year, the Fibonacci Association began publication of *The Fibonacci Quarterly*. In 1984 they began holding the regular international conferences focusing upon "Fibonacci numbers and their applications." The Fibonacci Association played a significant role in stimulating future international research.

Figure 1.15. Verner Emil Hoggatt, Jr. (1921–1980).

American mathematician Verner Emil Hoggatt, Jr. (1921–1980), professor at San Jose State University, was one of the founders of the Fibonacci Association and the magazine *The Fibonacci Quarterly*.

In 1969 Hoggatt published the book *Fibonacci and Lucas Numbers* [54], which is one of the best books in the field. Hoggatt made timely contributions to promote the research in the field of Fibonacci numbers. Scientific supervisor for many master's theses, Hoggat authored numerous articles on Fibonacci numbers.

Learned monk Brother Alfred Brousseau (1907–1988) was another prominent founder of the Fibonacci Association and *The Fibonacci Quarterly*.

The Fibonacci Association has the rather unique and singular purpose of studying mainly only the Fibonacci numbers and their generalizations. This raises some questions:

1. Why was the main purpose of the Fibonacci Association focused on Fibonacci numbers?
2. What united these two very different persons, the mathematician Hoggatt and the learned monk Brother Brousseau, in their quest to create the *Fibonacci Association* and to establish *The Fibonacci Quarterly*?

By attempting to answer these questions, we need to examine some of Hoggatt and Brousseau's documents, in particular, their photographs, as well as the books and articles, published in *The Fibonacci Quarterly*.

In 1969, TIME magazine published the article titled *The Fibonacci Numbers*, dedicated to the Fibonacci Association. This article contained Brousseau's photo with a cactus in his hands. The cactus is of course one of the most characteristic examples of the Fibonacci botanical objects. The article referred to other natural forms, involving Fibonacci numbers. For example, Fibonacci numbers are found in the spiral formations of sunflowers, pine cones, branching patterns of trees, and leaf arrangement (or phyllotaxis) on the branches of trees, etc.

Alfred Brousseau recommended to the lovers of Fibonacci numbers to "*pay attention to the search of aesthetic satisfaction in them. There is some kind of mystical connection between these numbers and the Universe.*"

We can see that Hoggatt holds a pine cone in his hands in photo (Fig. 1.15). The pine cone is another well-known example of the Fibonacci botanical objects, found in nature. From this comparison it may be reasonable to assume that Hoggatt, like Brousseau, believed in a mystical connection between Fibonacci numbers and the Universe. In the author's opinion, this belief is the main reason for Hoggat and Brousseau's primary motivation to study Fibonacci numbers and their manifestations in nature.

As indicated previously, Fibonacci numbers are linked with the golden ratio through Kepler's formula (1.20). This means that the Fibonacci numbers express the Harmony of Universe like the golden ratio, i.e., "*there is some kind of mystical connection between these numbers and the Universe*" (Alfred Brousseau).

This means that the theory of Fibonacci numbers, linked with the golden ratio, began to develop rapidly in modern science and this caused the creation of the Fibonacci Association (1963). The main goal of the Fibonacci Association was aimed primarily at solving applied problems of the harmonization of the theoretical natural sciences (physics, chemistry, botany, biology, physiology, medicine and so on), as well as economics, computer science, education and the Fine Arts. Thereby, the demonstration of the deep connection between Fibonacci numbers and nature is the best way to introduce the "harmonic ideas" of Pythagoras and Plato into many diverse fields of science.

### 1.10.3. *A role of Alan Turing in studying phyllotaxis phenomenon based on Fibonacci numbers*

Alan Turing (1912–1954) [62] is an English mathematician, logician, cryptographer, inventor of the *Turing machine*. Alan Turing has a broad and still growing reputation as one of the most creative thinkers of the 20th century. His research interests, from theoretical computer science to artificial intelligence and biology, covered many new research topics of the 21st century.

To celebrate the outstanding contribution of Alan Turing into theoretical computer science, the *Turing Award* had been established. The Turing Award is the most prestigious award in computer science, given by the Association for Computing Machinery (ACM) for outstanding scientific and technical contributions in the field of computer science.

Figure 1.16. Alan Turing (1912–1954).

Turing's achievements in computer science and technology are well known. Less widely known is that Turing was fond of solving scientific problem, which did not have, at first glance, direct relation to computer science. We are talking about phyllotaxis, well known botanical phenomenon, underlying the process of forming many botanical objects.

There is the question what relation Alan Turing had to phyllotaxis? After reading the article on Turing biography in Wikipedia [62], we find an unexpected answer to this question. In the Section "Pattern formation and mathematical biology" of [62] we read:

*"Towards the end of his life, Turing turned to mathematical biology, publishing "The Chemical Basis of Morphogenesis" in 1952. He was interested in morphogenesis, the development of patterns and shapes in biological organisms. His central interest in the field was understanding Fibonacci phyllotaxis, the existence of* ***Fibonacci numbers*** *in plant structures."*

When and where Turing showed interest in the problem of phyllotaxis? The answer to this question can be found in [63]. The article states that in secondary school Turing acquainted with the classic book *On Growth and Form*, published in 1917 by the English mathematician and biologist D'Arcy Wentworth Thompson (1860–1948). This book describes in details the phenomenon of phyllotaxis.

In Cambridge University (1947–1948), Turing attended lectures on physiology, and there he made the first attempt to give a logical description of the nervous system, and continued his studies in phyllotaxis. The first article of Turing on this subject [64] was published in 1952. Later, after his death in 1954, his followers edited and published in 1992 his second article on the subject [65].

Thus, the article [62] leads us to the following research achievements, which brought Turing the glory of the creator of theoretical computer science and a prominent scientist and thinker of the 20th century:

- *Turing machine*, which is an extension of a finite state machine, and according to the Church-Turing thesis is able to simulate any abstract computing machine that implements the process of step by step calculation.
- *Explanation of the cryptographic code of the German cipher machine "Enigma"*, what significantly affected the course of the Second World War.
- *Project of Automatic Computing Machine* ACE (Automatic Computing Engine).
- *Researches on phyllotaxis.*

Turing's researches on phyllotaxis can, of course, be regarded as Turing's *hobby*, which has no relation to the computer science and computer engineering. However, we should not forget that Alan Turing was one of the brilliant scientists and thinkers of the 20th century. And his researches on phyllotaxis, associated with his researches on the creation of logical model of the brain, a unique natural computer, can only be regarded as a brilliant prediction of the use of "Mathematics of Nature" and "Mathematics of Harmony" [4] for the creation of future computers.

Turing's researches in the field of phyllotaxis were continued by other scientists. Perhaps, the new geometry of phyllotaxis, developed by Ukrainian researcher Oleg Bodnar [59], is the highest modern scientific achievement in this area.

It follows from the above historical review that the phyllotaxis phenomenon led many mathematicians and researchers (including Verner Hoggatt and Alfred Brousseau) to Fibonacci numbers and their applications in nature. However, the prominent scientist and thinker Alan Turing was one of the first researchers in the world to understand the "*mystical connection between these numbers and the Universe*" and begun studying the phyllotaxis phenomenon, based on Fibonacci numbers! Turing's studies in phyllotaxis [63–65] had begun much earlier than the studies of American Fibonacci scientists and had a decisive influence on reviving the interest in Fibonacci numbers in modern science and led to the creation of the Fibonacci Association and *The Fibonacci Quarterly*. It is clear that Turing's research in phyllotaxis and later the researches of the members of Fibonacci Association led to a revival in modern science Pythagorean MATEM of *harmonics*. This idea underlies the author's speech "*The Golden Section and Modern Harmony Mathematics*," made at the 7th International Conference "*Fibonacci Numbers and Their Applications*" (Austria, Graz, July 1996) [25], which led the author to the publication of the fundamental book [4].

# Chapter 2

# A New View on Numeral Systems: Unusual Hypotheses, Surprising Properties and Applications

## 2.1. Dodecahedron, solar system, Egyptian calendar and Babylonian numeral system

In the period of the mathematics origin one of the "key" mathematical discoveries had been made. We are talking about the *Babylonian Positional Principle of Numbers Representation*. It is emphasized in [2] that "*the Babylonian sexagecimal numeral system, which arose approximately in 2000 BC, was the first numeral system, based on the positional principle*". It is necessary to note that the *Babylonian* Positional Principle underlie all well-known positional numeral systems, in particular, the widely used *decimal system* and *binary system*, the basis of modern computers.

There are different hypotheses about the origin of positional principle of number representation, described in the works [2, 66, 67]. However, it remains unclear why the Babylonians chose the number 60 as the base of their numeral system. An interesting hypothesis is described in [4]. This hypothesis has astronomical character and consists in the fact that the Babylonian and the Mayan's numeral systems are closely related to the Egyptian and the Mayan calendars, which in turn are related to the numerical characteristics of the dodecahedron and icosahedron — the main Platonic solids, which expressed the Harmony of the Universe in Plato's cosmology. Let us consider this hypothesis in more detail.

### 2.1.1. *A structure of the Egyptian calendar*

The Russian proverb says: "A time is the eye of history". Everything that exists in the Universe: the Sun, the Earth, asters, planets, known and unknown worlds, alive and non-alive nature — everything has time-spatial measurement. The time is measured by observation of the periodically repeating process of definite duration. The modern *Gregorian calendar* is the most widely used calendar. It was named after Pope Gregory XIII, who introduced it in October 1582. The Gregorian calendar dates back to the Egyptian calendar [68].

The ancient *Egyptian calendar* was a solar calendar with a 365-day year. This calendar, built in the 4th millennium BC, was one of the first solar calendars. In this calendar the year consisted of **365** days. The *year* was divided into **12** months, each *month* consisted of **30** days; at the end of the year the **5** holidays, which were not going into the month structure, had been added. Thus, the Egyptian calendar year had the following structure: **365 = 12 × 30 + 5**.

There is a question: why did Egyptians divide the calendar year into **12** months? We know that there were calendars with other number of months in the year. For example, in the Mayan's calendar the year has 18 months, each month has 20 days. The next question: why each month in the Egyptian calendar had exactly 30 days? One may ask some questions to the modern *UTC Time standard*, which divides the day into **24** hours, each hour into **60** minutes and each minute into **60** seconds. In particular, there is a question: why the unit of hour was chosen so that **1 day = 24 (2 × 12) hours**? Further: why **1 hour = 60 minutes**, and **1 minute = 60 seconds**? The same questions concern the choice of the measurement units of angular values, in particular: why a circumference is divided into 360°, that is, why **2π = 360° = 12 × 30°**? We can add other questions to these questions, in particular: why the astronomers recognized expedient to consider **12** "zodiacal" constellations, though actually during the motion on the ecliptic the Sun intersects 13 constellations? And further we can ask one "strange" question: *why the Babylonian numeral system had a rather exotic base, the number 60?*

### 2.1.2. *Connection of the Egyptian calendar with solar system and the numerical characteristics of the dodecahedron*

By analyzing the above questions, we find, that in them with the surprising constancy the following four numbers are repeated: **12**, **30**, **60** and the derivative from the number **360** (**360** = **12** × **30**). There is a question: is there some fundamental scientific idea, which could give simple and logic explanation of using these numbers in the Egyptian calendar, and in modern systems of the time and angle values measurement? To answer this question, we will return once again to the regular dodecahedron based on the golden ratio (Fig. 2.1).

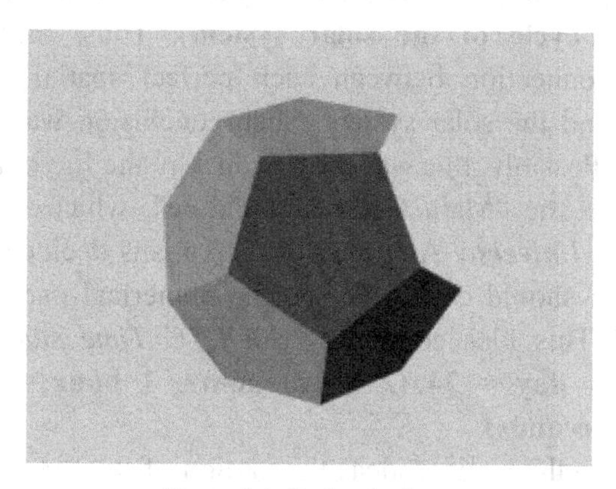

Figure 2.1. Dodecahedron.

Did the Egyptians know the dodecahedron? Historians of mathematics recognize that the ancient Egyptians knew the regular polyhedra. The ancient Greek philosopher and mathematician Proclus attributes to Pythagoras a geometric construction of all 5 regular polyhedra. But we know that Pythagoras borrowed at the ancient Egyptians many mathematical theorems and discoveries, in particular, Pythagoras' Theorem. From some data, Pythagoras had lived in Egypt 22 years and in Babylon 12 years. Therefore we can assume that Pythagoras also knew about the regular polyhedra from the ancient Egypt and Babylon.

But there are also other, more substantial proofs that the Egyptians possessed the information about all the 5 regular polyhedra. In particular, the playing dice, having the form of the icosahedron, that is, of "Platonic

Solid" dual to the dodecahedron can be seen in British Museum. From these facts we can put forward a hypothesis that the dodecahedron was known to the Egyptians. A much unexpected theory of the origin of the Egyptian calendar, and also of the modern systems of the time and geometric angles measurement follows from this hypothesis.

As it is known, the dodecahedron (Fig. 2.1) has **12** faces, **30** edges and **60** flat angles on its surface. If we accept the hypothesis that the ancient Egyptians knew the dodecahedron and its numerical parameters **12, 30, 60**, then what surprise the archaeologists was, when did they find out, that the cycles of the solar system are expressed by the same numbers (**12-year's cycle of Jupiter, 30-year's cycle of Saturn and, at last, 60-year's cycle of the solar system**). Thus, there is a deep mathematical connection between such perfect spatial figure as the dodecahedron and the solar system. Such conclusion was made by the archaeologists. Possibly, this was the reason why the Egyptians chose the dodecahedron as the "Main Geometric Figure", which symbolizes the *Harmony of the Universe*. And then the Egyptians decided that all their calendar system should correspond to the numerical parameters of the dodecahedron! This idea coincides with *UTC Time standard* of time measurement (**1 day = 24 (2 × 12) hours, 1 hour = 60 minutes, 1 minute = 60 seconds**).

According to the ancient ideas the motion of the Sun on the ecliptic had strictly circular nature, then by choosing the 12 Zodiac constellations with the distance in 30°, the Egyptians did coordinate surprisingly the year motion of the Sun on the ecliptic with the structure of their calendar year: *one month corresponded to the moving of the Sun on the ecliptic between two adjacent Zodiac constellations*! *Moreover, the movement of the Sun at the ecliptic on one degree did correspond to one day in the Egyptian calendar*! Thus, the ecliptic was divided automatically into 360°. Having divided a day into two parts, the Egyptians did divide then each half of one day into 12 parts (12 faces of the dodecahedron) and did introduce an *hour*, a major unit of a time. Having divided one hour into 60 minutes (60 plane angles on the surface of the dodecahedron), the chronologists introduced a *minute*, the next important unit of a time. In particular, they also introduced a *second* (*1 minute = 60 seconds*).

Thus, having chosen the dodecahedron as the "*Main Harmonic Figure of the Universe*" and strictly following to its numerical characteristics **12**, **30**, **60**, the Egyptians designed the perfect calendar. The Egyptian calendar completely corresponded to the "Theory of Harmony," based on the golden proportion, because this proportion underlies the dodecahedron.

Thus, such surprising conclusions follow from the comparison of the dodecahedron with the solar system.

And if our hypothesis is correct (let somebody try to deny it), it follows from here that *a lot of millennia the mankind lives under the flag of the golden section!* And each time, when we look at our watch, which are based on the numerical parameters of the dodecahedron 12, 30 and 60, we are dealing with the "*Main Secret of the Universe*", the *golden section*!

The next unusual hypothesis follows from these reasoning's. If we assume that the choice of base 60 of Babylonian numeral system has an astronomical origin, then this is an additional confirmation of our "*dodecahedral*" *hypothesis*, which establishes a connection between the 60-year cycle of the solar system, the Egyptian calendar, the base of Babylonian positional numeral system and the dodecahedron as the main Platonic Solid, which expresses the Harmony of the Universe, and the golden ratio, the main part of dodecahedron.

## 2.2. Icosahedron and its connection with solar system, Mayan's calendar and Mayan's numeral system

The appearance of the positional numeral systems is one of the major milestones in the history of material culture. Many nations have participated in their creation. In the 6th century AD the Mayan's created the original positional numeral system. The most widely belief is that the base of the Mayan's numeral system is the number 20 and therefore the Mayan system has a "finger" origin.

However, it is known that the Mayan's system has one deviation from the base **20** [2]. The weight of the next "node number," following after the number 20, was chosen in the Mayan's system to be **360** (instead of $\mathbf{20^2 = 400}$). This fact points to the astronomical origin of the

Mayan's numeral system. This fact is described in [2] as follows: "*This is the only deviation from the twenty-fold principle of the Mayan's system. It is explained by the fact that the Mayan's year was divided into 18 months by 20 days each plus five additional days.*"

According to [2], the following "node" numbers were the digit weights in the Mayan's positional system: $\{1, 20, 20 \times 18, 20^2 \times 18, 20^3 \times 18, \ldots\}$ that is, all the "node" numbers had been calculated by using the numbers 20 and 18, according to the following rule:

$$20^i \times 18 \ (i = 0, 1, 2, 3, \ldots), \tag{2.1}$$

where **18** is the number of months in Mayan's calendar.

By comparing the Platonic solids of *dodecahedron* and *icosahedron* with the Egyptian's and Mayan's calendars, and the Babylonian and Mayan's numeral systems, we found the following numerical coincidences.

Let us begin from *icosahedron* (Fig. 2.2). The *icosahedron* is Platonic solid, dual to *dodecahedron* (Fig. 2.1). It has **20** faces, **30** edges and **60** flat angles on its surface.

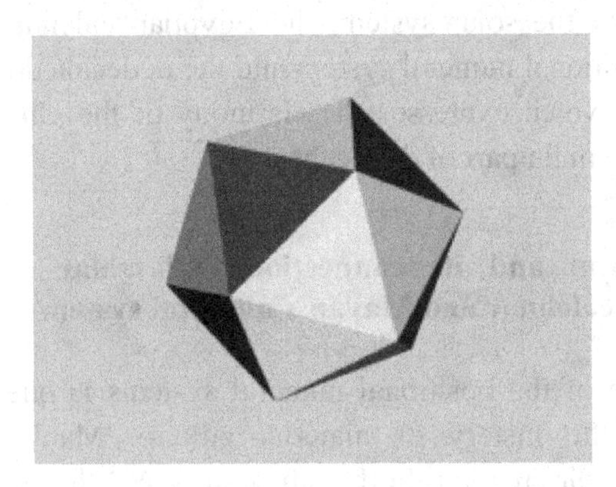

Figure 2.2. Icosahedron.

The Mayan used this calendar for highly accurate astronomical observation of the stars and the planets. The Mayan's calendar had the following structure: **1 year = 360 + 5 = 20 × 18 + 5**, that is, the Mayan's calendar year was divided into 18 months, every month has 20 days plus five additional days at the end. It is clear that the structure of the Mayan's calendar was similar to the Egyptian solar calendar: **1 year =**

**360 + 5 = 12 × 30 + 5**, based on the *dodecahedron*. Let us compare again the *icosahedron* (Fig. 2.2) and the *dodecahedron* (Fig. 2.1). The number of the icosahedron faces is equal to **20** and the number of the dodecahedron vertices is also equal to **20**! Both geometric figures are based on the *golden ratio*. Thus, the Mayan's undoubtedly used these numerical characteristics of the icosahedron in their calendar by means of the division of 1 year into **18** months.

The next important Mayan's idea was to connect their calendar with their positional numeral system. This was done by choosing the unusual values for the weights of digits of their positional numeral system ("node" numbers), given by (2.1). As the result, the Mayan's got the original system of mathematical knowledge coherent to astronomical knowledge. The golden section, which underlies icosahedron, united these mathematical and astronomical knowledge!

## 2.3. Surprising properties of the Egyptian decimal arithmetic

### 2.3.1. *Representation of numbers in the Egyptian decimal non-positional numeral system*

The ancient Egyptian decimal system emerged in the second half of the third millennium BC. It was decimal, but non-positional. The numbers have been represented as the sum of some "node numbers" 1, 10, $10^2$, $10^3$, $10^4$, $10^5$, $10^6$, $10^7$. In order to designate the "node numbers," Egyptians had used special hieroglyphic signs, which played the role of "node numbers". For example, for the number 325, Egyptians used 5 hieroglyphs, depicting the "node number" 1, 2 hieroglyphs, depicting the number 10, and 3 hieroglyphs, depicting the number 100.

### 2.3.2. *Method of doubling in Egyptian arithmetic*

The so-called *method of doubling* [67, 69] is considered the main achievement of the Egyptian arithmetic. With this method, Egyptians performed *multiplication* and *division* of numbers. In order to multiply 35 × 12, Egyptian mathematician performed the following. He constructed the table below. In the first column of the table he placed the binary numbers 1, 2, 4, 8, ..., $2^k$ ($k$ = 0, 1, 2, 3, ...); in the second column he placed the first number 35, and then in subsequent row of this column he placed the numbers equal to twice more of the previous row.

| | | | |
|---|---|---|---|
| 1 | 35 | | |
| 2 | 70 | | |
| **/4** | **140** | → | **140** |
| **/8** | **280** | → | **280** |
| | | | **35×12 = 420** |

Then he marked by inclined line those binary numbers with the sum equal to the second multiplier (12 = 4 + 8). The result of multiplication was obtained by adding the numbers, corresponding to the marked rows (140 + 280 = 35 × 12 = 420).

Analysis of the Egyptian method of multiplication, based on the "method of doubling," leads us to much unexpected conclusion. Decomposition of the second multiplier 12 by the sum of the powers of the binary numbers ($12 = 2^2 + 2^3$), used in the first column, is the same as representating it in the binary system (12 = 1100). On the other hand, if the multiplier 35 will be represented in the binary system (35 = 100011), then the used in the second column "doubling" of 35, carried out at each step of the multiplication, can be performed by shifting the binary code of the multiplier 35 by one bit leftward (70 = 1000110, 140 = 10001100 and so on).

In other words, the above Egyptian method of multiplication, based on the "doubling method," is essentially the same as the basic algorithm of multiplication of numbers in modern computers!

As it is shown in [67, 69], the division is performed by using the *method of doubling* and this method of division is similar to the algorithm of the division, used in modern computers!

Thus, about four thousand years ago the ancient Egyptian mathematicians made the great mathematical discovery for future computer technology; they invented the *method of doubling*, which is the basis of the most important arithmetical algorithms of multiplication and division, used in today's computers!

Below we explain how this mathematical discovery can be used in the *Fibonacci arithmetic*, which lies at the basis of the *Fibonacci computers*, the new direction in computer technology.

This example is a convincing proof of the fact that it is important to explore the history of ancient mathematics and science so as to understand the development of modern computer science and technology.

## 2.4. Decimal system

### 2.4.1. *A history of the decimal system*

The oldest known recording of positional decimal system was discovered in India in 6th century AD. The introduction of *zero* is the greatest achievement of the Indians. As it is highlighted in [2], "*the introduction of zero was absolutely unavoidable stage of the natural development that led to the creation of contemporary positional system.*"

The *zero* was unknown for the Egyptians, the Romans, Greeks and the Hebrews. Only in India, the *zero* finally took its place in mathematics and then the *zero* had spread around the world. Indian numeral system came first to the Arab countries and then to Western Europe. The famous mathematician al-Khwarizmi from Central Asia talked about this numeral system in his works. Simple and easy rules of addition and subtraction of numbers, written in a positional system, made it especially popular. As the work of al-Khwarizmi was written in Arabic, the Indian numeration was named in Europe "*Arab numeral system*" or "*Arab numerals*" (though it was incorrectly from a historical point of view).

The earliest manuscript in Arabic, containing the Indian positional numeral system, was published in 9th century. French clergyman and mathematician Herbert, who in 999 became a Roman pope under the name Sylvester II, was one of the first who understood the advantages of decimal numeration. The new Pope Sylvester II tried to reform the teaching of mathematics on the basis of new numeration. Unfortunately, the innovation met furious anger of the Inquisition. The Pope-mathematician was accused that he "sold his soul to Saracen devils." The reform had been ruined and the Pope-mathematician died.

Although the first recordings in the Arab-Indian numeral system were found in the Spanish manuscripts in the 10th century, the decimal system began to take root in Europe only in the 12th century. New numeration in Europe met with resistance from both the scholastic science, and the governments. For example, in 1299 in Florence, merchants were not allowed to use the Arab numerals; only Roman numerals in the accounts or to write the numbers by words were allowed.

There are many opinions about the choice of the number 10 as the base of the decimal system; the most common is the opinion that the number 10 has "finger" origin. However, we should not forget that the number 10 always had a special meaning in ancient science. The Pythagoreans called this number *tetractys*. The *tetractys* 10 = 1 + 2 + 3 + 4 was considered by the Pythagoreans as one of the highest values and was the "*symbol of the Universe*," because it contained the four *key elements*: *monad* =1, indicated by Pythagoras as the *spirit*, from which the whole visible world derives; *deuce* or *dyad* (2 = 1 + 1), symbolizing the *material atom*; *triple* or *triad* (3 = 2 + 1), the symbol of the *living world*. *Four* or *quadruple* (4 = 3 + 1) connects the *living world* with the *monad* and thus symbolized the *whole*, that is, the *visible and invisible*. Thus, the *tetractys* 10 = 1 + 2 + 3 + 4 expressed *Everything*. Thus, the *harmonious* hypothesis, stet the origin of the number 10, is no less right to exist as the "*finger*" one.

### 2.4.2. *The greatest mathematical discovery in mathematics history*

We widely use *decimal system* for everyday's calculations. This positional numeral system is the most common worldwide. Everybody can agree with the statement that any child after graduation from the fourth grade of secondary school should know at least two useful things: to read and write on native language and to use decimal system to perform elementary arithmetic operations. The *decimal numeral system* is one of the most important achievements of human intellect. This numeral system is based on the *Babylonian Positional Principle*. The decimal system seems to be such simple and elementary, that it is difficult to agree with the statement that the *decimal system is one of the greatest mathematical discoveries in the history of mathematics*.

To prove the validity of this statement let us turn to the opinion of the "authorities" [2].

Pierre-Simon Laplace (1749–1827), the Great French mathematician, member of the Parisian academy of sciences, an honorable foreign member of the Petersburg academy of sciences:

*"The idea of representation of all numbers by using 9 marks, giving to them, apart from value by the form, another value by the place too, seems so simple what namely because of this simplicity it is difficult to understanding as this is surprising. As not easy to come to this method, we see on the example of the greatest geniuses of Greek science Archimedes and Apollonius, from whom this idea remained latent."*

M.V. Ostrogradsky (1801–1862), the Great Russian mathematician, a member of the Petersburg academy of sciences and other foreign academies:

*"It seems to us that after the invention of written language the largest discovery was the use by humanity of the so-called decimal numeral system. We want to say that the agreement, with the aid of which we can represent all useful numbers by twelve words and by their endings is one of the most remarkable creations of human genius..."*

Jules Tannery (1848–1910), the French mathematician, a member of the Parisian academy of sciences:

*"As to the present system of written numeration in which we use the nine significant numerals and a zero, and the relative value of numerals is defined by a special rule, this system has been introduced in India into the epoch which is not determined precisely, but, apparently, after the Christian era. The invention of this system is one of the most important events in history of science, and despite a habit to use decimal numeration, we should be surprised by extraordinary simplicity of its mechanism."*

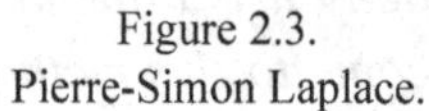

Figure 2.3.
Pierre-Simon Laplace.

Figure 2.4.
M.V. Ostrogradsky.

Figure 2.5.
Jules Tannery.

## 2.5. History of binary system: since ancient time to John von Neumann's Principles

### 2.5.1. *The origin of the binary system*

The binary system is considered one of the oldest positional numeral systems. Many nations and outstanding scholars participated in its creation. We briefly describe the main stages in the creation of the binary system and binary arithmetic, as it is described in the Wikipedia [69].

We have mentioned that the Egyptian *method of doubling* is closely related to binary numbers and used for ancient Egyptian multiplication and division. This method was used, for instance, in the *Rhind Mathematical Papyrus*, which dates to around 1650 BC.

The famous Chinese *Book of Changes* is the most significant work of ancient Chinese philosophy. The *Book of Changes* was published in 9th century BC. It consists of 64 characters — *hexagrams*, each of which expresses a particular life-time situation from the point of view of its gradual development. The full set of 8 trigrams and 64 hexagrams, which are essentially analogs of 3-bit and 6-bit binary numbers, was known in ancient China from the classic texts of the *Book of Changes*. The hexagrams in the *Book of Changes* were arranged in accordance with the values of the corresponding binary digits (0 to 63), and a method to obtain these was developed by the Chinese scholar and philosopher

Shao Yong. However, there isn't evidence to show that Shao Yong has developed the rules for binary arithmetic.

The next mention of the binary system is available in the works of Indian poet and mathematician Pingala (200 BC), who developed a mathematical foundation for describing poetry by binary system. He used binary numbers in the form of short and long syllables (the latter equal in length to two short syllables), making it similar to Morse code.

### 2.5.2. *Leibniz's binary arithmetic and the "Book of Changes"*

The prominent German scientist Gottfried Wilhelm Leibniz (1646–1716) created the binary arithmetic.

Figure 2.6. Gottfried Wilhelm Leibniz.

Since he was a student until his death, he studied the properties of the binary system, which in the future has become the basis of modern computers. The binary system has been fully described by Leibniz in XVII century in the work "*Explanation of Binary Arithmetic, which uses only the characters 1 and 0, with some remarks on its usefulness, and on the light it throws on the ancient Chinese figures of Fu Xī*" (1703).

Leibniz attributed to the binary system a mystical meaning, and believed that by using it, we can create a universal language to explain the phenomena of the world.

As Leibniz was interested in Chinese culture, he was aware of "*Book of Changes*" and one of the first to notice that the hexagram corresponds to the binary numbers from 0 to 111111. Leibniz believed that the "*Book of Changes*" is evidence of major Chinese accomplishments in mathematical philosophy of the time.

Leibniz did not recommend the binary system instead of a decimal to practical calculations, but stressed that "*the calculation by using deuces 0 and 1, in rewarding to its length, is major in science, and even in the calculation practice, especially in geometry: the reason is the fact that at the reduction of numbers to the simplest principles, what are the 0 and 1, everywhere the wonderful order could be established.*" [2]. By this statement, Leibniz anticipated the modern "computer revolution," based on the binary system!

### 2.5.3. *George Boole and Claude Shannon*

In 1854, British mathematician George Boole (1815–1864) published the article, detailing an algebraic system of logic, which is now known as Boolean algebra. His logical calculus became the important instrumental in designing digital electronic circuitry.

In 1937, American expert in computer science Claude Shannon (1916–2001) produced his master's thesis at MIT, which implemented Boolean algebra and binary arithmetic in electronic relays and switches for the first time in science history. Entitled *A Symbolic Analysis of Relay and Switching Circuits*, Shannon's thesis was essentially the first practical guidebook on digital circuit design.

Figure 2.7. George Boole.

Figure 2.8. Claude Shannon.

### 2.5.4. *Development of new concepts in designing computers*

In 1942 in Pennsylvania University under Mauchly and Ekkert's supervision, the works on designing the computer (ENIAC) were started and completed at the end of 1945. In February 1946, the first public demonstration of the (ENIAC) was held. A role of the ENIAC in the development of computer science is determined first of all by the fact that the ENIAC was the first acting machine, in which electronic elements were used for the fulfilment of arithmetical and logic operations and also for storage of information. In this machine the use of the new electronic technology allowed one to increase the speed of computer approximately in 1000 times in comparison to electromechanical computers.

A direct outcome of the machine ENIAC was efficiency of electronic technology in computers. It was necessary to analyze the pros and cons of ENIAC project and to give appropriate recommendations. A brilliant solution of this task was given in the famous Report "*Preliminary discussion of the logical design of an electronic computing instrument*" (1946) [70]. This Report, written by the brilliant mathematician John von Neumann and his colleagues from the Prinstone Institute for Advanced Study Goldstein and Berks, did present the project of new electronic computer.

Figure 2.9. John von Neumann (1903–1957).

The essence of the main recommendations of this *Report*, named *John von Neumann's Principles*, was the following:

(1) The machines on electronic elements should work not in the decimal system but in the **binary system**.
(2) The program should be placed in the machine block, called *storage device*, which should have a sufficient capacity and appropriate speed for access and entry of program commands.
(3) Programs, as well as numbers, with which the machine operates, should be represented in binary code. Thus, the commands and the numbers should have one and the same form of representation. This meant that *the programs and all intermediate outcomes of calculations, constants and other numbers should be placed in the same storage device.*
(4) The difficulties of physical realization of the storage device, speed of which should correspond to the speed of logical elements, do demand on *hierarchical organization of memory.*
(5) The arithmetical device of the machine should be constructed on the basis of the logical summation element; it is inadvisable to create special devices for the fulfilment of other arithmetical operations.
(6) The machine should use *parallel principle* of the organization of computing processes, that is, the operations over the binary words should be fulfilled over all digits simultaneously.

Thus, the historical significance of *John von Neumann's Principles* consists of the fact that they are a brilliant confirmation of Leibniz's predictions about the role of binary system in the future development of commuter science and technology. The prominent American scientist, physicist and mathematician John von Neumann (1903–1957), together with his colleagues gave *strong preference to the binary system as the universal way of coding of data in electronic computers.*

Neumann's idea to use binary system in electronic computers is based on arithmetic advantages of the binary system and specifics of electronic components. Von Neumann wrote:

*"Our main memory unit by nature is adapted to binary system ... A trigger in fact is again a binary device ... The main advantage of the binary system in comparison with a decimal consists in greater simplicity*

*of technical realization and big speed, with which the basic operations can be performed.*

*An additional remark consists of the following. The main part of the computer by its nature is not arithmetical, but mainly logical. The new logic, being the system "yes-no" is mainly binary. Therefore, the construction of the binary arithmetical devices greatly facilitates the construction of more homogeneous machine, which can be designed better and much effectively."*

Thus, as it is emphasized by many outstanding mathematicians, the discovery of the *positional principle of number representation* (Babylon), and then of *decimal system* (India), as well as the creation of the *binary system* and *binary arithmetic* (Egypt, China, India, Leibniz and others) can be classified as truly epoch-making mathematical discoveries, which influenced greatly on the development of material culture, in particular, the development of computer science and technology.

## 2.6. Canonical and symmetrical numeral systems

### 2.6.1. *The concept of canonical positional numeral systems*

The above-mentioned *Babylonian numeral system* with base 60, *decimal* and *binary* systems belong to the class of the so-called *canonical positional numeral systems* [6], which are defined by the following general expression:

$$x=\sum_{i=0}^{n-1} b_i R^i, \tag{2.2}$$

where $x$ is a real number, $R$ is the base of numeral system (2.2), $b_i$ is the numeral of the $i$th digit, and $R^i$ is the weight of the $i$th digit. The numerals of the numeral system (2.2) take their values from the alphabet $A = \{a_1, a_2, \ldots, a_m\}$. For the decimal system the alphabet $A = \{0, 1, 2, 3, 4, 5, 6, 7, 8, 9\}$, for the binary system $A = \{0, 1\}$.

The binary system with the alphabet $A = \{0, 1\}$

$$x = \sum_{i=0}^{n-1} b_i 2^i, \tag{2.3}$$

is the simplest among *canonical positional numeral systems* (2.2).

The abridged designation of the numeral system (2.2) has the following form:

$$x = b_n b_{n-1} \ldots b_0, b_{-1} b_{-2} \ldots b_{-k}. \tag{2.4}$$

Conventionally, the base of the canonical numeral system (2.2) is natural number greater than or equal to 2. In this case, the feature of the canonical numeral system (2.2) is the fact that the base $R$ of the numeral system (2.2) coincides with the number $m$ of the numerals in the alphabet $A = \{a_1, a_2, \ldots, a_m\}$, that is, $R = m$.

However, sometimes another definition of the base of numeral system is used, when the base is equal to the ratio of the adjacent digit weights. For example, the bases of the decimal and binary systems are equal, respectively:

$$10 = \frac{10^{i+1}}{10^i} \text{ and } 2 = \frac{2^{i+1}}{2^i}. \tag{2.5}$$

In conclusion of this section, we should do one important remark. Because the set of positive integers from 2 to infinity can be chosen as the bases of the canonical numeral system (2.2), this means that the number of canonical numeral systems (2.2) is theoretically infinite. However, only the numeral systems with the bases 2, 3, 10, 12, 20, 60 have certain practical applications. All other canonical systems, given by (2.2) and have no practical implementation, are of purely theoretical significance.

We note that the choice of numeral system for practical application is determined primarily by the requirements of *simplicity* of technical realization. In this respect, the binary system is brilliant example. *A simplicity of technical implementation of the binary system in electronic computers was the main reason to choose the binary system as a basis of modern computer technology.*

### 2.6.2. *The conception of symmetrical numeral systems*

The *symmetrical numeral systems* [6] are a generalization of the concept of the canonical positional numeral systems (2.2); this generalization is realized by expanding the concept of the "numeral" in the positional representation. As it is known, the numerical equivalent of the numerals in the canonical numeral system (2.2), as a rule, are integers, which take values from the set {0, 1, 2, …, $R - 1$}, where $R$ is the base of numeral system.

The main disadvantage of the numeral system (2.2) is the impossibility of representing negative numbers without introducing a special "sign bit." As it is known, for representation of negative numbers in computers, the concepts of the *inverse* and *additional* codes are widely used. Thus, the negative numbers in the canonical positional numeral systems (2.2) play the role of certain "rogues," which cannot be represented in canonical numeral systems (2.2) in the "direct" code.

In order to eliminate this disadvantage, the so-called *symmetrical numeral systems* have been introduced [6]. The symmetrical numeral systems are given by the following sum:

$$N = \sum_{i=0}^{n-1} b_i R^i \tag{2.6}$$

where $R = 2S + 1$ is the base of numeral system (2.6), $b_i \in \left\{\overline{S}, \overline{S-1}, \ldots, \bar{1}, 0, 1, 2, \ldots, S\right\}$, $S = -S$, $n$ is the number of digits in the system (2.6).

Thus, the symmetrical numeral system (2.6) has two important features:

1. The base of the system (2.6) is always the odd number $R = 2S + 1$, that is, $R = 3, 5, 7, \ldots$ .
2. The numerals of the symmetrical numeral system take the values from the set $b_i \in \left\{\overline{S}, \overline{S-1}, \ldots, \bar{1}, 0, 1, 2, \ldots, S\right\}$, where $\overline{S} = -S$. This means that the symmetrical system (2.6) uses $2S + 1$ numerals; one of these numerals has the numeric value of 0, the remaining $2S$ numerals are divided into two groups: the first of them has a positive numerical values 1, 2, 3, …, $S$; the second group has negative numerical values

$\bar{1},\bar{2},\bar{3},...,\bar{S}$. Such choice of the numerals has significant advantages in comparison with the canonical positional numeral system (2.2), because the symmetrical system (2.6) can represent all negative numbers without introducing the concept of the *inverse* and *additional* codes. It is easy to prove [74] that with the $n$ digits we can represent in the symmetrical system (3.5) $R^n$ of integers (positive and negative, including the number of 0) in the range from

$$N_{\min} = -\frac{R^n - 1}{2} \text{ to } N_{\max} = \frac{R^n - 1}{2}. \tag{2.7}$$

## 2.7. Ternary-symmetrical system and ternary arithmetic

### 2.7.1. *Ternary-symmetrical representation*

As follows from the definition (2.6), the number of symmetrical numeral system (2.6) is theoretically infinite. However, the *ternary-symmetrical numeral system* is of the most practical significance; it is given by the following sum:

$$N = \sum_{i=0}^{n-1} b_i 3^i, \tag{2.8}$$

where $b_i \in \{\bar{1},0,1\}$ is the ternary numeral of $i$th digit, $3^i$ is the weight of the $i$th digit, $n$ is the number of digits in the system (2.8).

The ternary-symmetrical numeral system (2.8), which has base 3 and uses three numerals $\{\bar{1},0,1\}$ for number representation, is the simplest among the symmetrical systems (2.6), that is, *the requirement of the simplicity of technical implementation is the main reason for choosing this system for computer technology.*

Clearly, the abridged notations of the maximum and minimum numbers in the system (2.8) can be represented as follows, respectively:

$$N_{\max} = \underbrace{111...1}_{n} \tag{2.9}$$

$$N_{\max} = \underbrace{\bar{1}\bar{1}\bar{1}...\bar{1}}_{n}. \tag{2.10}$$

Ternary representations (2.9), (2.10) have the following algebraic interpretation, respectively:

$$N_{\max} = 3^{n-1} + 3^{n-2} + ... + 3^1 + 3^0 = \frac{3^n - 1}{2} \tag{2.11}$$

$$N_{\min} = -3^{n-1} - 3^{n-2} - ... - 3^1 - 3^0 = -\frac{3^n - 1}{2}. \tag{2.12}$$

Then it follows from the above arguments the following theorem [6].
**Theorem 2.1.** *By using n ternary digits, we can represent in the ternary-symmetrical numeral system (2.8)* $3^n$ *integers (including* $\frac{3^n - 1}{2}$ *positive numbers,* $\frac{3^n - 1}{2}$ *negative numbers and the number of 0) in the range from* $N_{\min} = -\frac{3^n - 1}{2}$ *to* $N_{\max} = \frac{3^n - 1}{2}$.

For example, by using 3-digit ternary-symmetrical code, we can represent $3^3 = 27$ (in general $3^n$) integers (including 13 positive, 13 negative numbers and the number 0) in the range from $N_{\min} = -\frac{3^3 - 1}{2} = -13$ (in general $N_{\min} = -\frac{3^n - 1}{2}$) to $N_{\max} = \frac{3^3 - 1}{2} = 13$ (in general $N_{\max} = \frac{3^n - 1}{2}$).

### 2.7.2. *Representation of negative numbers*

Let us consider the two $n$-digit numbers $A$ and $B$, represented in the ternary-symmetrical system (2.8):

$$A = 1b_{n-2}b_{n-3}...b_1b_0 \tag{2.13}$$

$$B = \bar{1}b_{n-2}b_{n-3}...b_1b_0. \tag{2.14}$$

Note that the number $A$ has the positive numeral 1, and the number $B$ has the negative numeral $\bar{1}$ in the most significant digits of the ternary-symmetrical representations (2.13) and (2.14).

It is easy to show that for arbitrary values of the rest numerals $b_{n-2}\, b_{n-3} \ldots b_1\, b_0$ in (2.13), the ternary code combination (2.13) represents positive number only, regardless of the values of the rest digits in the codeword $b_{n-2}\, b_{n-3} \ldots b_1\, b_0$. This conclusion follows immediately from the inequality:

$$3^{n-1} > \frac{3^{n-1}-1}{2}, \tag{2.15}$$

where $3^{n-1}$ is the weight of the most significant, that is, $(n-1)$th digit in (2.13) and $\frac{3^{n-1}-1}{2} = \underbrace{11...1}_{n-1}$ is the maximum sum of the rest digit weights in the code word $b_{n-2}b_{n-3} \ldots b_1 b_0$ in the ternary-symmetrical representation (2.13).

For the case (2.14), the inequality (2.15) turns into another inequality:

$$-3^{n-1} < -\frac{3^{n-1}-1}{2}. \tag{2.16}$$

It follows from (2.16) that the ternary-symmetrical code (2.14) always represents negative number, regardless of the values of the other digits of the codeword $b_{n-2}\, b_{n-3} \ldots b_1\, b_0$ in (2.14).

It follows from this reasoning a very important conclusion that the sign of the number, presented in the ternary-symmetrical system (2.8), is contained in the most significant digit 1 or $\bar{1}$ of the ternary-symmetrical representations (2.13) or (2.14); in this case, if the most significant digit is positive bit of 1, then the number is positive, in opposite case the number is negative.

Thus, the ternary-symmetrical system (2.8) does not require a special symbol to represent the sign of number. Positive and negative numbers are represented in the "direct" code. This important property is invariant in arithmetic operations. This means that all arithmetic operations in system (2.8) can be performed in the "direct" code without using the concepts of the *inverse* and *additional* codes.

It should be noted that the important advantage of the ternary-symmetrical system (2.8) is the existence of the very simple rule for obtaining the number of opposite sign from the ternary-symmetrical representation of the initial numeral. For this case, we are applying the original ternary-symmetrical representation the rule of *ternary inversion*, the essence of which is as follows:

$$\bar{1} \to 1;\ 0 \to 0;\ 1 \to \bar{1}. \tag{2.17}$$

**Example 2.1.** Let the initial ternary-symmetrical representation of the decimal number 56 has the following form:

$$56_{10} = 1\bar{1}01\bar{1}_3.$$

Applying (2.17) to this ternary-symmetrical representation, it is easy to get the ternary-symmetrical representation of the number with opposite sign, that is,

$$-56_{10} = \bar{1}10\bar{1}1_3.$$

### 2.7.3. *Ternary-symmetrical summation and subtraction*

The following elementary identity for the powers of base 3 underlies the ternary-symmetrical summation:

$$3^i + 3^i = 3^{i+1} - 3^i. \tag{2.18}$$

The next table of the ternary-symmetrical summation follows from this identity.

Table 2.1. Ternary-symmetrical summation.

| $a_k + b_k$ | $\bar{1}$ | 0 | 1 |
|---|---|---|---|
| $\bar{1}$ | $\mathbf{\bar{1}1}$ | $\bar{1}$ | 0 |
| 0 | $\bar{1}$ | 0 | 1 |
| 1 | 0 | 1 | $\mathbf{1\bar{1}}$ |

A number of peculiarities of the ternary-symmetrical summation follow from Table 2.1. These peculiarities appear at the summation of the ternary numerals of the same sign, namely:

$$\mathbf{1+1=1\bar{1}} \text{ and } \mathbf{\bar{1}+\bar{1}=\bar{1}1}.$$

We can see that there appears the intermediate sum and carry-over at the summation of the ternary numerals of the same sign. In this case the sign of the carry-over coincides with the sign of the summable numerals; however, the sign of the intermediate sum is opposite.

The ternary-symmetrical subtraction of the numbers $A-B$ is reduced to the summation of the numbers $A + (-B)$, if the rule of the ternary inversion (2.17) is applied to the subtrahend $(-B)$.

**Example 2.2.** Adding two ternary-symmetrical numbers $64_{10} = 1\bar{1}101$ and $16_{10} = 1\bar{1}\bar{1}1$.

**Solution:**

$$\begin{array}{rccccc} & 1 & \bar{1} & 1 & 0 & 1 \\ + & & & & & \\ & & 1 & \bar{1} & \bar{1} & 1 \\ \hline & 1 & 0 & 0 & 0 & \bar{1} \end{array}$$

The summation result $1000\bar{1} = 1\times 3^4 + \bar{1}\times 3^0 = 81-1 = 80$ is a positive number, because its ternary-symmetrical representation begins with the positive numeral 1.

**Example 2.3.** Subtract the ternary-symmetrical number $16_{10} = 1\bar{1}\bar{1}1$ from the ternary-symmetrical number $64_{10} = 1\bar{1}101$.

**Solution.** Subtraction of two ternary numbers 64−16 is reduced to the summation of the numbers 64+(−16), if we apply to the negative number (−16) the rule of the ternary inversion (2.17):

$$(-16) = \bar{1}11\bar{1}.$$

Then, the subtraction is reduced to the summation:

$$
\begin{array}{r}
1\ \bar{1}\ 1\ 0\ 1 \\
+ \\
\bar{1}\ 1\ 1\ \bar{1} \\
\hline
1\ \bar{1}\ \bar{1}\ 1\ 0
\end{array}.
$$

The subtraction result $1\bar{1}\bar{1}10 = 1\times3^4 + \bar{1}\times3^3 + \bar{1}\times3^2 + 1\times3^0 = 81 - 27 - 9 + 1 = 46$ is a positive number, because its ternary-symmetrical representation begins with the positive numeral 1.

### 2.7.4. *Ternary-symmetrical multiplication*

The ternary-symmetrical multiplication (Table 2.2) is based on the following trivial mathematical identity for the number 3 powers:

$$3^m \times 3^n = 3^{m+n}. \tag{2.19}$$

It follows from (2.19) the following Table 2.2 for the ternary-symmetrical multiplication.

Table 2.2. Ternary-symmetrical multiplication.

| $a_k \times b_k$ | $\bar{1}$ | 0 | 1 |
|---|---|---|---|
| $\bar{1}$ | 1 | 0 | $\bar{1}$ |
| 0 | 0 | 0 | 0 |
| 1 | $\bar{1}$ | 0 | 1 |

**Example 2.4.** Multiply two ternary-symmetrical numbers $(-10)_{10} = \bar{1}0\bar{1}$ and $2_{10} = 1\bar{1}$.

**Solution:**

$$
\begin{array}{r}
\bar{1}\ 0\ \bar{1} \\
\times \\
1\ \bar{1} \\
\hline
1\ 0\ 1 \\
\bar{1}\ 0\ \bar{1}\ \ \\
\hline
\bar{1}\ 1\ \bar{1}\ 1
\end{array}.
$$

You can see that the ternary-symmetrical multiplication is reduced to the ternary-symmetrical summation of two partial products, that are formed as the result of the multiplication of the first multiplier $\bar{1}0\bar{1}$ by the lowest ternary numeral $\bar{1}$ of the second multiplier and then by the highest ternary numeral 1 of the second multiplier.

Note that we have multiplied the negative number $(-10)_{10}$ by the positive number $2_{10}$ in the "direct" code. After summation, we obtained the following result of multiplication:

$$\bar{1}1\bar{1}1 = \bar{1}\times 3^3 + 1\times 3^2 + \bar{1}\times 3^1 + 1\times 3^0 = -27+9-3+1 = -20.$$

By looking at the product of ternary-symmetrical multiplication $(-20)_{10} = \bar{1}1\bar{1}1$, we can see that the product is a negative number, because its ternary-symmetrical representation $\bar{1}1\bar{1}1$ begins with the negative numeral $\bar{1}$.

### 2.7.5. *Ternary-symmetrical division*

The ternary-symmetrical division is reduced to sequential left-shift of the divisor, when the highest significant numeral (1 or $\bar{1}$) of the shifted divisor will coincide with the highest significant numeral of the dividend. Let us consider the case, when the divisor has been shifted on the $k$ digits to the left. Then, the shifted divisor is compared with the dividend. If the signs of the highest significant numerals of the shifted divisor and the dividend coincide, then the highest significant numeral of the first partial quotient $Q_1$ is assumed to be 1 and the first partial quotient takes the form:

$$Q_1 = 1\underbrace{00\ldots0}_{k}, \tag{2.20}$$

where the number of 0's after 1 is equal to $k$. For this case the shifted divisor is subtracted from the dividend.

If the highest significant numerals of the shifted divisor and the dividend are opposite by sign, then the first partial quotient $Q_1$ is assumed to be $\bar{1}$ and the first partial quotient takes the form:

$$Q_1 = \bar{1}\underbrace{0\,0\ldots 0}_{k}, \qquad (2.21)$$

where the number of 0's after $\bar{1}$ is equal to $k$. For this case the shifted divisor is summarized with the dividend.

As a result of the first stage of the mirror-symmetrical division, we obtain the first partial quotient in the form (2.20) or (2.21) and the first intermediate dividend $D_1$ as the result of summation or subtraction of the shifted divisor from the dividend.

The next stage of the ternary-symmetrical division consists in comparing the first intermediate dividend $D_1$ with the shifted divisor according to the rules, described above.

The procedure of the comparison of the intermediate dividend and the shifted divisor continues until we obtain the intermediate dividend, equal to 0, or when the exactness of the division result becomes acceptable to us. Then, we have to sum up all partial quotients to obtain the result of ternary-symmetrical division.

### 2.7.6. *The basic functions of the ternary logic and their technical realization*

The ternary logic is a special case of the so-called $k$-valued logic ($k = 2, 3, 4, 5, \ldots$) for the case $k = 3$. For the coordination with the ternary-symmetrical numeral system (2.8) we will assume that the ternary logical variables take their values from the set $\{\bar{1}, 0, 1\}$.

Then the basic logical functions of *one ternary variable* v are determined as follows:

| *Inversion function* | *Cyclic negation* |
|---|---|
| $f(v) = \bar{v} = \begin{cases} \bar{1} \text{ with } v = 1 \\ 0 \text{ with } v = 0 \\ 1 \text{ with } v = \bar{1} \end{cases}$ | $f(v) = \overset{\approx}{v} = \begin{cases} \bar{1} \text{ with } v = 0 \\ 0 \text{ with } v = 1 \\ 1 \text{ with } v = \bar{1} \end{cases}$ |

Let us consider the following important functions of *two ternary variables*:

(1) *Ternary conjunction* $f(v_1, v_2) = min(v_1, v_2) = v_1 \wedge v_2$

| $\wedge$ | $\bar{1}$ | 0 | 1 |
|---|---|---|---|
| $\bar{1}$ | $\bar{1}$ | $\bar{1}$ | $\bar{1}$ |
| 0 | $\bar{1}$ | 0 | 0 |
| 1 | $\bar{1}$ | 0 | 1 |

(2) *Ternary disjunction* $f(v_1, v_2) = max(v_1, v_2) = v_1 \vee v_2$

| $\wedge$ | $\bar{1}$ | 0 | 1 |
|---|---|---|---|
| $\bar{1}$ | $\bar{1}$ | 0 | 1 |
| 0 | 0 | 0 | 1 |
| 1 | 1 | 1 | 1 |

(3) *Addition by modulo 3* $f(v_1, v_2) = v_1 \oplus v_2 \ (mod\ 3)$

| $\oplus$ | $\bar{1}$ | 0 | 1 |
|---|---|---|---|
| $\bar{1}$ | 1 | $\bar{1}$ | 0 |
| 0 | $\bar{1}$ | 0 | 1 |
| 1 | 0 | 1 | $\bar{1}$ |

(4) *Multiplication by modulo 3* $f(v_1, v_2) = v_1 \otimes v_2 \ (mod\ 3)$

| $\otimes$ | $\bar{1}$ | 0 | 1 |
|---|---|---|---|
| $\bar{1}$ | 1 | 0 | $\bar{1}$ |
| 0 | 0 | 0 | 0 |
| 1 | $\bar{1}$ | 0 | 1 |

There are the following identities for the above ternary logical functions:

$$\bar{\bar{v}} = v; \ v \wedge v = v; \ v \wedge \bar{1} = \bar{1}; \ v \wedge 1 = v;$$

$$v \vee v = v; \ \ v \vee 1 = 1; \ \ v \vee \bar{1} = v;$$

$$v \oplus 0 = v;\ \ v \otimes 0 = 0;\ \ v \otimes 1 = v;\ \ v \otimes \bar{1} = \bar{v}.$$

The ternary functions of *conjunction, disjunction* and *inversion* are related by Morgan's formulas:

$$\overline{v_1 \wedge v_2} = \bar{v}_1 \vee \bar{v}_2;\quad \overline{v_1 \vee v_2} = \bar{v}_1 \wedge \bar{v}_2.$$

Note that the three-valued logic appeared for engineers as the long known logic of the positive, negative and zero electrical current, and for programmers as the logic of number signs: +, 0, −, etc.

Similarly to the Boolean logic there are different variants of the functionally complete systems of the ternary logic functions. We can use the functions of the so-called "modular logic" to synthesize the ternary logic elements. The system of the "modular logic" includes the following functions:

$$f(v_1, v_2) = v_1 \oplus v_2;\ \ f(v_1, v_2) = v_1 \otimes v_2. \tag{2.22}$$

We can add to the modular functions (2.22) a special function

$$f(v_1, v_2) = v_1 \Theta v_2. \tag{2.23}$$

The function (2.23) gives a rule of the carry-over formation at the summation of the single-digit ternary-symmetrical numbers. The logic table of this function has the following form:

| $\Theta$ | $\bar{1}$ | 0 | 1 |
|---|---|---|---|
| $\bar{1}$ | $\bar{1}$ | 0 | 0 |
| 0 | 0 | 0 | 0 |
| 1 | 0 | 0 | 1 |

## 2.7.7. *Technical realization of the ternary logic functions*

The set of the ternary logic functions (2.22), (2.23) are the functionally complete set of the ternary logic functions, which may be used for synthesizing the ternary logic elements. It is easy to prove that the ternary inversion function $\bar{v}$ and the cyclic negation function $\tilde{\tilde{v}}$ are realized by using the modulo 3 addition logic elements (Fig. 2.10).

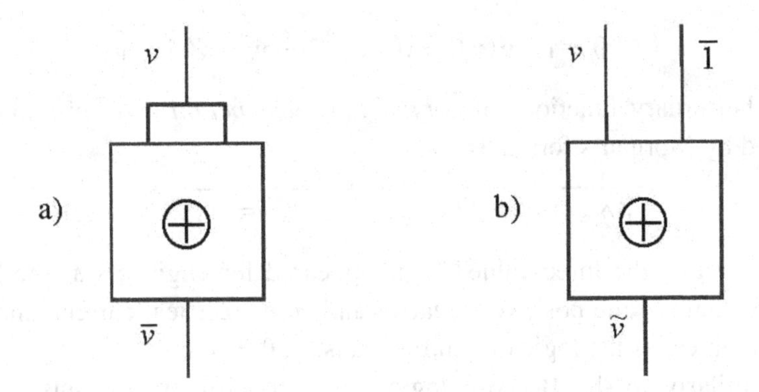


Figure 2.10. The logic elements of the ternary inversion (a) and the cyclic negation (b).

The ternary single-digit half-summator of the kind 2Σ is realized by using the logic elements ⊕ and Θ (Fig. 2.11(a)) and the ternary single-digit multiplier is based on the logic element ⊗ (Fig. 2.11(b)).

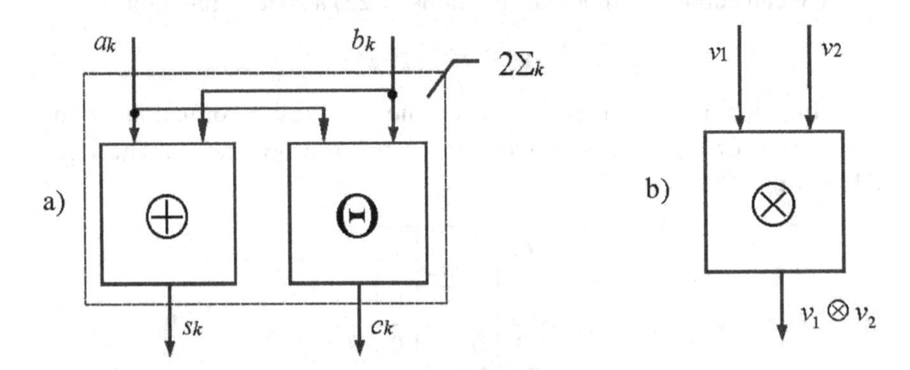


Figure 2.11. The ternary single-digit half-summator (a) and multiplier (b).

If we take the ternary single-digit half-adder of the kind 2Σ as the basic logic element for the design of the ternary-symmetrical arithmetical devices, we can prove that the 5 single-digit half-summators form the ternary single-digit full summator of the kind 4Σ (Fig. 2.12).

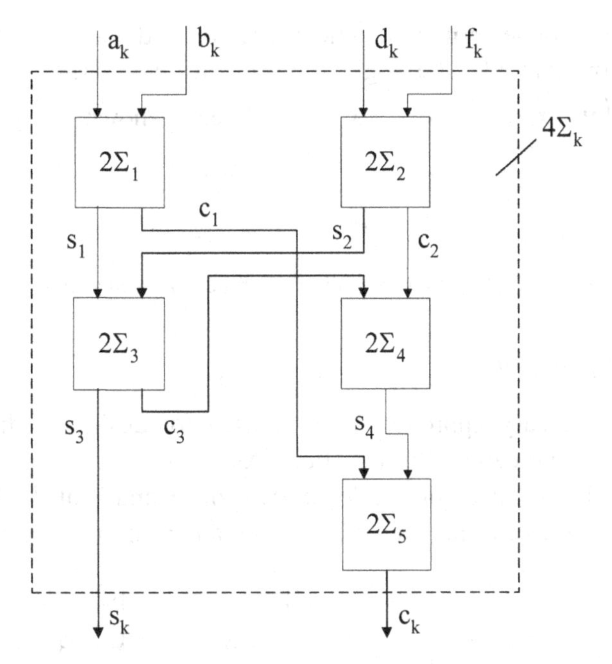


Figure 2.12. The full ternary single-digit summator.

### 2.7.8. *The binary realization of the ternary logic elements*

For the micro-electronic realization by using VLSI, we can use the following binary coding of the ternary variables as shown in Table 2.3.

Table 2.3. Binary coding ($x_1$, $x_2$) of the ternary numerals $v$.

| v | $x_1$ | $x_2$ |
|---|---|---|
| $\bar{1}$ | 1 | 0 |
| 0 | 0 | 0 |
| 1 | 0 | 1 |

Using Table 2.3, each ternary numeral can be represented by means of VLSI with the binary code word $x_1x_2$. Then the problem of designing ternary logic elements is reduced to designing binary VLSI.

Note that some ternary functions are realized very simply in this manner. For example the logic element of the "ternary inversion" $f(\nu)=\bar{\nu}\left(\nu=x_1x_2 \text{ and } \bar{\nu}=x_2x_1\right)$ is realized as shown in Fig. 2.13.

Figure 2.13. The binary realization of the *ternary inversion.*

### 2.7.9. *Flip-flap-flop*

The similar "binary approach" can be used for designing the ternary memory element called flip-flap-flop. As is well known, the classic binary flip-flop is based on the logic elements 1 and 2 of the kind OR-NOT (Fig. 2.14(a)), which are connected by the back logic connections.

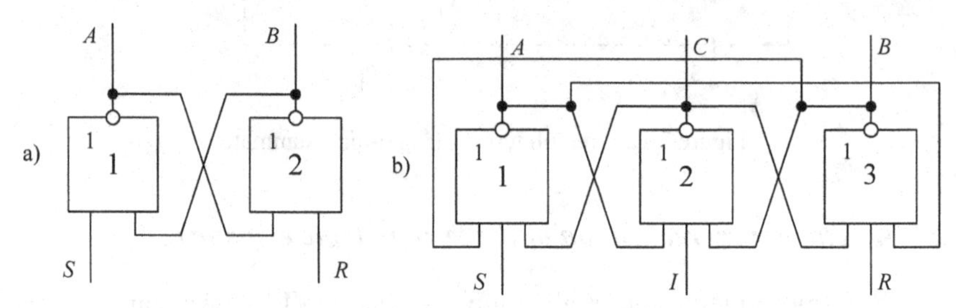


Figure 2.14. "Flip-flop" (a) and "flip-flap-flop" (b).

Let us consider now the logic circuit, which consists of the three logic elements 1, 2, 3 of the kind *OR-NOT* (Fig. 2.14(b)). Suppose that the logic elements 2 and 3 are adjacent to the logic element 1, the logic elements 3 and 1 are adjacent to the logic element 2, and the logic elements 1 and 2 are adjacent to the logic element 3. Each logic element *OR-NOT* is connected with its adjacent logic elements by the back logic connections. This is a cause of the three stable states of the logic circuit in Fig. 2.14(b). In fact, suppose that we have the logic 1 on the input *C* of the logic element 2. This logic 1 enters the inputs of the adjacent logic elements 2 and 3 and supports the logic 0 on their outputs *A* and *B*. These logic 0's enter the inputs of the logic element 2 and support the logic 1 on its output *C*. Hence, this state of the circuit in Fig. 2.14(b) is the first stable state. This stable state corresponds to the code combination 0 1 0 on the outputs $A$, $C$, $B$. One may show that the circuit has two stable states corresponding to the code combinations 1 0 0 and 0 0 1 on the outputs $A$, $C$, $B$. In fact, it is easy to show that the logic 1 on the output $A$ is a cause of the second stable state 100 of the logic circuit in Fig. 2.14(b). At least, the logic 1 on the output $B$ is a cause of the third stable state of the logic circuit in Fig. 2.14(b). We can use the above-mentioned stable states of the circuit in Fig. 2.14(b) for the binary coding of the ternary numerals according to the following table:

$$0 = 010$$

$$1 = 001$$

$$\bar{1} = 100$$

If we eliminate the middle output $C$ we will get the binary outputs $A$ and $B$, which correspond to the binary coding of the ternary variables according to Table 2.2.

Hence, the logic circuit in Fig. 2.14(b) can be considered as the ternary-binary memory element called *flip-flap-flop*. Let us consider now the functioning of the "flip-flap-flop" in Fig. 2.14(b). It has three stable states $\bar{1}$, 0 and 1. Let the "flip-flap-flop" in Fig. 2.14(b) be in the state $Q = 0$. This means that the output $C = 1$, and other outputs $A = B = 0$. If we need to set the "flip-flap-flop" into the state $Q = 1$ (0 0 1) we have to send to the "flip-flap-flop" inputs $S$, $I$, $R$ the following adjusting signals $S = 1$, $I = 1$, $R = 0$. The signals $S = 1$ and $I = 1$ cause the appearance of the logic 0's on the outputs $A$ and $C$. These logic 0's enter the inputs of the logic element 3 and together with the logic signal $R = 0$ cause an appearance of the logic 1 on the output $B$.

By analogy one may show that the adjusting signals $S = 0$, $I = 1$, $R = 1$ turn the "flip-flap-flop" in Fig. 2.14(b) into the state $\bar{1}$ (100).

## 2.8. Brusentsov's ternary principle and ternary technology

### 2.8.1. *Nikolay Brusentsov (1925–2014): first steps in computer engineering*

Nikolay Brusentsov was born in 1925 in Kamianske, Ukraine. In 1953 he graduated from the Moscow Energy Institute and begun to work at the

Special Design Bureau (SDB), Moscow University as a computer engineer. In the beginning of his career Brusentsov had a chance to participate in the adjustment of the M-2 computer, which had been constructed in the special computer laboratory of the Soviet Academy of Sciences under the supervision of the famous Soviet computer expert M. Kartsev. The M-2 computer, which was one of the best Soviet computers of that period, literally won Brusentsov's heart and he began to dream about the construction of his own computer. His intention coincided with the decision of the famous Soviet mathematician academician Sergey Sobolev, who headed the Computational Mathematics Department, Moscow University, to create new computer for educational purposes in Moscow University. Sergey Sobolev organized the scientific seminar, in which the Moscow mathematicians and programmers Shura-Bura, Semendjaev, Jogolev and Sobolev himself participated. They discussed the deficiencies of today's binary computers and considered their architecture and various technical methods of their design. After these discussions the seminar decided to give a preference to magnetic elements as element-base of new computer. The preferable basic elements at that period were magnetic cores and diodes because transistors then did not exist but vacuum tubes were excluded because of their low reliability and their large size and dissipation.

Figure 2.15. Nikolay Brusentsov (1925–2014).

### 2.8.2. *First steps to the ternary computer "Setun"*

Intensive work on the creation of the new computer began in April 1956, when Sobolev formulated the project of the new computer at the scientific seminar. Brusentsov was appointed as the principal developer of the project. According to Sobolev's initiative, for realization of this project later the *Problem Computer Laboratory,* Moscow University was organized at the Mechanics and Mathematics Faculty and Nikolay Brusentsov was appointed as a Head of this Laboratory in 1962. The first idea was to construct a traditional binary computer, based on the magnetic elements. After studying the projects of the binary magnetic computer, Brusentsov had found many shortcomings in them. By that time he decided to elaborate the ternary magnetic computer. In one of his articles Brusentsov wrote: "*Of course, I knew the advantages of the ternary code from the special books, which devoted much attention to it. Later I have found that the American scientist Grosch ("Grosch's law") was interested in the ternary number system but the American scientists did not create the ternary computer.*"

In 1957 Brusentsov designed the basic devices of the ternary computers: summator, counter and other typical devices. In 1958 the engineers of the laboratory made the first specimen of the ternary computer and the computer began to function within 10 days since the beginning of the complex debugging! It was named "Setun" after the name of the small river Setun not far from the Moscow University.

### 2.8.3. *The basic specifications of the computer "Setun"*

The computer "Setun" was one-addressed fixed-point computer of sequential functioning. From functional point of view the computer was divided into six devices: arithmetic device, control device, device for operative memory, input device, output device, memory device on the magnetic drum. From the mathematical point of view a peculiarity of the "Setun" computer was the use of the *ternary symmetrical numeral system* with the ternary numerals $\{\bar{1},0,1\}$. From the engineering point of view a peculiarity of the computer was using the magnetic pulse amplifier as the basic element of the computer. Such amplifier consisted of non-linear transformer with miniature magnetic core and germanium diode. Three stable states, which are necessary for the ternary representation, were realized by means of the use of a pair of such amplifiers. The 18-digits ternary code combination was the "ternary word." In the arithmetic device of the computer "Setun" the 18-digit ternary word was considered as a number, in which the point is located between the second and third digits. The commands were coded by the half-word, that is, by the 9 ternary digits.

### 2.8.4. *Implementations of the computer "Setun"*

According to the Resolution of the Soviet Ministerial Council, the full-scale production of the computer "Setun" was assigned to the Kazan computer factory. The Kazan computer factory had produced 50 specimens of the computer "Setun". A majority of them (30 specimens) was delivered to the Soviet universities. The computers "Setun" functioned productively in all climate zones from Kaliningrad to Magadan, and from Odessa and Ashgabat to Novosibirsk. The computers functioned reliably practically without any service and without spares.

### 2.8.5. *The dramatic fate of the computer "Setun"*

During 1961–1968 Nikolay Brusentsov designed the architecture of the new ternary computer named "Setun-70". The algorithm of its functioning was described on the programming language "ALGOL". Unfortunately, Brusentsov's laboratory after designing the computer of "Setun-70" was deprived of a possibility to develop the ternary computers because of the negative attitude of the Moscow University administration to his computer ideas. Brusentsov's laboratory was deported to the attic of the student dormitory. According to the decision of the University administration the experimental specimen of "Setun," which worked without failures for 17 years, was destroyed barbarously. The computer was bisected for parts and was thrown to the garbage dump!

### 2.8.6. *Brusentsov's ternary principle*

It is well known that the design of computers begins with the choice of numeral system, which determines many technical characteristics of computers. At the dawn of the computer era the choice of the "optimal" number system for electronic computers was solved brilliantly by the outstanding American physicist and mathematician John von Neumann, which gave emphatic preference to the binary system in electronic computers. The famous *John von Neumann's Principles* include three basic ideas of the electronic computer design: *binary numeral system, binary (Boolean) logic, binary memory element ("flip-flop")*.

However, researches in the field of numeral systems continued after designing the first binary computers. The basic motivations of these researches are connected by overcoming essential lacks the classic binary system. It was considered that the binary system has two serious shortcomings. The *sign problem* is the first of them. Its essence consists in the fact that it is impossible to represent negative numbers and perform arithmetical operations over them in the "direct" binary code which complicates arithmetical computer structures. The *problem of "zero" redundancy* is the second serious shortcoming of the binary system. As all binary code combinations are "allowed", this fact complicates the problem of checking errors, which can appear in computers in process of storage of information, its transmission and data processing.

The first real attempt to overcome the first shortcoming of the binary system (the *sign problem*) was made in the Soviet science at the dawn of the computer era. The original computer project (the ternary "Setun" computer) [71], designed in 1958 in Moscow University (the Principal Designer Nikolay Brusentsov), was a clever example of the "optimal" solution of the "sign problem". A new principle of the computer design was realized in the "Setun" computer of Moscow University. This principle was based on the following ideas: *ternary logic, ternary-symmetric numeral system, ternary memory element ("flip-flap-flop")*. This principle is now called in modern computer science *Brusentsov's Ternary Principle* [28]. Let us compare now the *binary digital technology* based on *Neumann's binary principle* and the *ternary digital technology* based on *Brusentsov's ternary principle*. The binary digital technology is based on the two-valued signals (*bits*) and two-stable memory elements (*flip-flop*). The discrete objects, having more than 2 values, are represented as combinations of bits or bites (8 bits). For example, decimal numerals are represented by the "fours" of bits, symbols of alphabet by bites, etc. Accordingly, all operations over the not two-valued objects are realized as consequences of the operations of the two-valued logic.

The ternary digital technology is based on the three-valued signals (*trits*) and three-stable memory elements (*flip-flop-flap*). The objects, having more than 3 values, are represented as combinations of trits. The operations over these objects are realized as the consequences of the operations of the three-valued logic. The analog of the "bite" (8 bits) is the combination of the 6 trits, called a *trite*.

One of the barriers restraining the development and spreading the ternary digital technology is misunderstanding about extraordinary nature and hard comprehensibility of the three-valued logic. In reality the three-valued logic is not only quite correct and adequate to the reality but is more convenient and habitual for people's form of thinking than the two-valued logic. In real life we encounter three-valued relations very often. For example, "to increase — do not change — to decrease", "forward — stop — backwards", "*A* wins — a tie — *B* wins", "a surplus — a norm — a shortage", "friendly — neutral — hostile", "sooner — in time — later", "to the left — directly — to the right", etc.

Many modern computer scientists came to conclude that the ternary principle of computer design could become the alternative principle of future computer progress. In this connection it is pertinent to recall the opinion of the well-known Russian scientist Prof. Pospelov to the so-called *ternary-symmetrical numeral system*, used in Setun computer. In his book [6] he wrote the following:

*"The barriers, which stand on the way of the application of ternary-symmetrical numeral systems in computers, have the technical character. Up to now the economical and effective elements with three stable states have not been elaborated yet. As soon as such elements are developed, a majority of computers of the universal kind and many special computers*

*will most probably be designed so that they would work in the ternary-symmetrical numeral system."*

Also the famous American expert in computer science Prof. Donald Knuth [72] expressed the opinion that the replacement of "flip-flop" by "flip-flap-flop" would happen some day.

It should be noted that the development of processors, based on ternary numeral system, is actively continuing in modern science. For example, the publication [73] describes the ternary digital processor for performing a fast Fourier transform.

Furthermore, a group of scientists from the University of Pennsylvania has developed a nano-wire memory of high capacity, which is able to preserve the data, represented in ternary code, instead of the commonly used binary code [74].

The binary and ternary numeral systems are the most striking examples of how the numeral systems may affect the development of information technology. The binary system has already led to the emergence of modern information technology. There is every reason to believe that the ternary-symmetrical system may lead to the creation of a new "ternary information technology" or even "ternary information revolution" because ternary computers are the direction to go.

# Chapter 3

# Bergman's System, "Golden" Number Theory and Mirror-Symmetrical Arithmetic

## 3.1. George Bergman and Bergman's system

### 3.1.1. *Definition of Bergman's system*

In 1957 the young American mathematician George Bergman published the article *A number system with an irrational base* in *Mathematics Magazine* [10]. The following sum is called *Bergman's system*:

$$A = \sum_i a_i \Phi^i, \tag{3.1}$$

where $A$ is any real number, $a_i$ is a binary numeral $\{0, 1\}$ of the $i$th digit, $i = 0, \pm 1, \pm 2, \pm 3, \ldots$, $\Phi^i$ is the weight of the $i$th digit, and $\Phi = \left(1+\sqrt{5}\right)/2$ is the base of the numeral system (3.1).

### 3.1.2. *About George Bergman*

We can get more detailed information about George Bergman from Wikipedia [75]. We can read in [75]: "*George Mark Bergman was born on 22 July 1943 in Brooklyn, New York. He ... received his PhD from Harvard in 1968, under the direction of John Tate. The year before he had been appointed Assistant Professor of mathematics at the University of California, Berkeley, where he has taught ever since, being promoted to Associate Professor in 1974 and to Professor in 1978. His primary research area is algebra, in particular associative rings, universal algebra, category theory and the construction of counter-examples.*

*Mathematical logic is an additional research area. Bergman officially retired in 2009, but is still teaching.*"

Figure 3.1. George Bergman.

It is interesting to note the following. The concept of Bergman's system has been used widely in Internet and modern scientific literature. The special article in Wikipedia [76] is dedicated to *Bergman's system*. It is described briefly in WolframMathWorld [77]. Donald Knuth refers to Bergman's article [10] in his outstanding book [72]. The special paragraph in author's book [4] is dedicated to Bergman's system. "*The Computer Journal*" (British Computer Society) published in 2002 author's article [28]; this article is based on Bergman's system (4.1) and is dedicated to the so-called *ternary mirror-symmetrical arithmetic*, which was evaluated highly by Prof. Donald Knuth. Thus, the article [10] had glorified Bergman's name more than his other mathematical works, published in adulthood. It is surprising that there is no mention about Bergman's outstanding article [10] in Wikipedia Bergman's biographical article [75].

### 3.1.3. *The main distinction between Bergman's system and binary system*

On the face of it, there is no essential distinction between the formula (3.1) for Bergman's system and the formulas for the canonic positional numeral systems, in particular, *binary system*:

$$A=\sum_i a_i 2^i \left(i=0,\pm1,\pm2,\pm3,\ldots\right)\left(a_i\in\{0,1\}\right), \tag{3.2}$$

where the digit weights are connected by the following "arithmetical" relations:

$$2^i = 2^{i-1} + 2^{i-1} = 2 \times 2^{i-1}, \tag{3.3}$$

which underlie "binary arithmetic".

The principal distinction of *Bergman's system* (3.1) from the *binary system* (3.2) is the fact that the irrational number $\Phi = \left(1+\sqrt{5}\right)/2$ (the golden ratio) is used as the base of the numeral system (3.1) and its digit weights are connected by the following relations:

$$\Phi^i = \Phi^{i-1} + \Phi^{i-2} = \Phi \times \Phi^{i-1}, \tag{3.4}$$

which underlie the "golden" arithmetic.

That is why; Bergman called his numeral system the *numeral system with irrational base*. Although Bergman's article [10] is a fundamental result for number theory and computer science, mathematicians and experts of computer science of that period were not able to appreciate the mathematical discovery of the American wunderkind.

## 3.2. The "Golden" number theory and new properties of natural numbers

### 3.2.1. *Euclid's definition of natural numbers*

It is well known, that a number is one of the most important notions of mathematics and number theory is one of the most famous ancient mathematical theories.

But what is a number? It would seem for the first sight that mathematicians came to the common answer to this question. But all not so is simple. There are various definitions of the number. The simplest of them is used by Euclid in his *Elements*.

Euclid considered all numbers as geometric segments, and such geometric approach led him to the following geometric definition of natural numbers. Suppose that we have the infinite number of the "standard segments" of length 1. Euclid named them "monads" and he did not consider the "monads" as the numbers. It was simply the "beginning of all numbers." It is clear that for the construction of all natural numbers we should have the infinite set $S$ of the "monads", that is,

$$S = \{1, 1, 1, \ldots\}. \tag{3.5}$$

Then we can define a natural number $N$ as some geometric segment, which can be represented as the sum of the "monads" taken from (3.5), that is,

$$N = \underbrace{1 + 1 + 1 + \ldots + 1}_{N}. \tag{3.6}$$

In spite of limiting simplicity of definition (3.6), it had played a great role in mathematics, in particular in number theory. This definition underlies many important mathematical concepts, for example, concepts of the *prime* and *composed* numbers, and also the concept of *divisibility*, which is one of the main concepts of number theory.

### 3.2.2. *Constructive definition of real numbers*

But there are also other definitions of a number. The so-called "constructive approach" to the definition of "real number" is known, for example. According to this approach the real number $A$ is some mathematical object, which can be represented in the binary system (3.2). The definition of the real number $A$, given by (3.2), has the following geometric interpretation. Consider now an infinite set of the "binary" segments of length $2^n$, that is,

$$B = \{2^n\}(n = 0, \pm 1, \pm 2, \pm 3, \ldots). \tag{3.7}$$

Then all the real numbers can be represented by the sum (3.2), which consists of the "binary" segments, taken from (3.7).

Note that the number of the terms, including the finite sum (3.2), is always *finite* but *potentially unlimited*, that is, the definition (3.2) is a brilliant example of the *potential infinity concept*, used in "constructive" mathematics [78, 79].

Clearly, that the definition (3.2) sets on the numerical axis only a part of real numbers, which can be represented by the *finite* sum (3.2). We will name such numbers *constructive real numbers*. All other real numbers, which cannot be represented by the *finite* sum (3.2), are *non-constructive real numbers*.

What numbers can be referred to as "non-constructive" numbers within the framework of (3.2)? Clearly, all irrational numbers, in particular, the main mathematical constants $\pi$ and $e$, the number $\sqrt{2}$, the golden ratio are referred to as "non-constructive" numbers. But within the framework of (3.2) some "rational" numbers (for example, 2/3, 3/7, etc.), which cannot be represented by the *finite* sum (3.2), should be referred to as "non-constructive" numbers.

Note that though the definition (3.2) considerably limits the set of real numbers, this fact does not belittle its significance from the "practical" computing point of view. It is easy to prove that any "non-constructive" real number can be represented in the form (3.2) approximately, and the approximation error $\Delta$ will decrease as the terms in (3.2) increase, however $\Delta \neq 0$ for all the "non-constructive" real numbers. In essence, in modern computers we use only the "constructive" numbers, given by (3.2), though we do not have any problem with the "non-constructive" numbers, because they can be represented in the form (3.2) with the approximation error $\Delta \neq 0$, which strives to 0 potentially.

### 3.2.3. *Newton's definition of a real number*

In the 17th century in modern mathematics, a number of methods of studying the "continuous" processes have been developed and the concept of real number again appeared. Most clearly the new definition of this concept was given by Isaac Newton, one of the founders of mathematical analysis, in his *Arithmetica Universalis* (1707):

*"We understand a number not as the set of units, but as the abstract ratio of some magnitude to other of the same kind, taken as the unit."*

This formulation gives the universal definition of a real number, rational or irrational. If you consider now the *Euclid's definition* (3.6) from the point of *Newton's definition*, we can see that the "monad" in (3.6) plays a role of the unit. In the *binary system* (3.2) the number 2, that is, the base of *binary system*, plays a role of the unit.

### 3.2.4. *Again about the "extended" Fibonacci and Lucas numbers*

*Bergman's* system (3.1) is connected closely with the so-called *"extended" Fibonacci and Lucas numbers* $F_i$ and $L_i$ ($i$ = 0, ±1, ±2, ±3, ...), introduced in Chapter 1 (see Table 3.1).

Table 3.1. The "extended" Fibonacci and Lucas numbers.

| $n$ | 0 | 1 | 2 | 3 | 4 | 5 | 6 | 7 | 8 | 9 | 10 |
|---|---|---|---|---|---|---|---|---|---|---|---|
| $F_n$ | 0 | 1 | 1 | 2 | 3 | 5 | 8 | 13 | 21 | 34 | 55 |
| $F_{-n}$ | 0 | 1 | −1 | 2 | −3 | 5 | −8 | 13 | −21 | 34 | −55 |
| $L_n$ | 2 | 1 | 3 | 4 | 7 | 11 | 18 | 29 | 47 | 76 | 123 |
| $L_{-n}$ | 2 | −1 | 3 | −4 | 7 | −11 | 18 | −29 | 47 | −76 | 123 |

As it is shown in Chapter 1, the "extended" Fibonacci and Lucas numbers are connected by the following relations:

$$F_{-n} = (-1)^{n+1} F_n;\ L_{-n} = (-1)^n L_n. \tag{3.8}$$

### 3.2.5. *Bergman's system as a new definition of a real number*

In the above we developed the so-called "constructive" approach to the definition of real number, based on binary system (3.2). And this idea allows the following generalization. We can extend Newton's definition of real number for the case of *Bergman's system* (3.1). In fact, such interpretation of *Bergman's system* (3.1) has great theoretical importance for modern mathematics and its history. It changes our ideas about real numbers. Historically natural numbers were the first class of real numbers; the irrational numbers were introduced into mathematics much later, after the discovery of the "incommensurable segments." In the traditional numeral systems (Babylonian sexagecimal, decimal, binary) some natural numbers 60, 10, 2 are used as the "beginning of calculus". All real numbers can be represented with the help of the bases 60, 10 or 2. In *Bergman's system* (3.1) the golden ratio $\Phi = \left(1+\sqrt{5}\right)/2$ is the "beginning of calculus". All the other real numbers (including natural numbers) can be represented through the golden ratio $\Phi=\left(1+\sqrt{5}\right)/2$. This means that the irrational number $\Phi=\left(1+\sqrt{5}\right)/2$ (the golden ratio) is becoming the major number of mathematics, since it allows representing in *Bergman's system* (3.1) all numbers (natural, rational, irrational), as will be shown below.

### 3.2.6. *The "golden" representations of natural numbers*

A new definition of real numbers, based on *Bergman's system* (3.1), can be a source for new number-theoretical results. We begin our research from the "golden" representations of natural numbers in *Bergman's system* (3.1). With this purpose we will study the following representation of natural numbers $N$ in *Bergman's system* (3.1):

$$N=\sum_i a_i\Phi^i, \tag{3.9}$$

where $a_i \in \{0, 1\}$ is the bit of the $i$th digit, $\Phi^i$ is the weight of the $i$th digit, $\Phi=\frac{1+\sqrt{5}}{2}$ is the base of the numeral system (3.9).

We will name the sum (3.9) the *Φ-code of natural number N*. The abridged notation of the Φ-code of natural number $N$ has the following form:

$$N = a_n a_{n-1}\ldots a_1 a_0.a_{-1}a_{-2}\ldots a_{-k}. \tag{3.10}$$

and is named the *"golden" representation of natural number N*.

Note that the point in (3.10) separates the "golden" representation (3.10) into two parts: the left-hand part, where the bits $a_n a_{n-1}\ldots a_1 a_0$ have non-negative indices, and the right-hand part, where the bits $a_{-1}a_{-2}\ldots a_{-k}$ have negative indices.

Note that the weights $\Phi^i$ of the *Φ-code* (3.9) are connected by the following relation:

$$\Phi^i=\Phi^{i-1}+\Phi^{i-2}. \tag{3.11}$$

Besides, the power of the golden ratio $\Phi^i$ is expressed through the "extended" Fibonacci and Lucas numbers (see Table 3.1) as follows:

$$\Phi^i = \frac{L_i + F_i\sqrt{5}}{2} (i = 0, \pm 1, \pm 2, \pm 3, ...). \tag{3.12}$$

Let us apply now the rule

$$N' = N + 1 \tag{3.13}$$

for obtaining all natural numbers.

In order to apply the rule (3.13) for obtaining all the "golden" representations of natural numbers in the form (3.10), we need to transform the "golden" representation (3.10) of the initial number $N$ to such form, when the bit of the 0th digit will become equal to 0, i.e. $a_0 = 0$. We can fulfill such transformation by means of the micro-operations of *convolution* and *devolution*, based on the fundamental identity (3.11).

$$\textit{Convolution}: 011 \rightarrow 100 \tag{3.14}$$

$$\textit{Devolution}: 100 \rightarrow 011 \tag{3.15}$$

If we add the binary 1 to the 0th digit of the *"golden" representation* (3.10), we realize the rule (3.13).

We begin a demonstration of this method from the transformation of the "golden" representation of number 1 to the "golden" representation of number 2. The "golden" representation of number 1 in the form (3.10) has the following form:

$$1 = 1.0 = 1 \times \Phi^0 + 0 \times \Phi^{-1} = \Phi^0. \tag{3.16}$$

By using the micro-operation of the *devolution* (3.15), we get other "golden" representation of number 1 as follows:

$$1 = 0.11 = \Phi^{-1} + \Phi^{-2}. \tag{3.17}$$

After that we can apply the rule (3.13) to the "golden" representation (3.17), that is, we add the bit of 1 to the 0th digit of the "golden" representation (3.17). As a result, we get the "golden" representation of number 2:

$$2 = 1.11 = \Phi^0 + \Phi^{-1} + \Phi^{-2}. \tag{3.18}$$

If we apply the operation of "convolution" (3.14) to the "golden" representation (3.18), we get other "golden" representation of number 2:

$$2 = 10.01 = \Phi^1 + \Phi^{-2}. \tag{3.19}$$

By adding the bit of 1 to the 0th digit of the "golden" representation (3.19) and by fulfilling the *convolution* (3.14), we get the following "golden" representation of number 3:

$$3 = 11.01 = 100.01 = \Phi^2 + \Phi^{-2}. \tag{3.20}$$

The "golden" representation of number 4 follows from (3.20):

$$4 = 101.01 = \Phi^2 + \Phi^0 + \Phi^{-2}. \tag{3.21}$$

We get the "golden" representation of number 5 from the "golden" representation (3.21), if we fulfill the following transformation of (3.21) by using the *devolutions*:

$$4 = 101.01 = 101.0011 = 100.1111. \tag{3.22}$$

By adding the bit of 1 to the 0th digit of the right-hand "golden" representation (3.22), we get the following "golden" representation of number 5:

$$5 = 101.1111. \tag{3.23}$$

By performing the *convolutions* in the "golden" representation (3.23), we get the new "golden" representations of number 5:

$$5 = 101.1111 = 110.0111 = 1000.1001 = \Phi^3 + \Phi^{-1} + \Phi^{-4}. \tag{3.24}$$

By continuing this process *ad infinitum*, we can get the "golden" representations of all natural numbers. Thus this consideration leads to the following unexpected mathematical result, which can be formulated in the form of the following theorem.

**Theorem 3.1.** *All natural numbers can be represented in the Φ-code (3.9) of Bergman's system (3.1) by the finite number of bits.*

We note that Theorem 3.1 is far from trivial, if we take into consideration that all the powers of the golden proportion $\Phi^i$ ($i = \pm1, \pm2, \pm3, \ldots$) in (3.9) (with the exception of $\Phi^0 = 1$) are irrational numbers. Hence, it follows from Theorem 3.1 that any natural number can be represented as a finite sum of the powers of the golden proportion $\Phi^i$ ($i = \pm1, \pm2, \pm3, \ldots$), that is, arbitrary natural number is *constructive* in the framework of the definition of (3.1) .

Note that Theorem 3.1 is true for all natural numbers. Therefore, Theorem 3.1 can be referred to the category of new properties of natural numbers.

### 3.2.7. *Multiplicity and minimal form of the "golden" representations*

As follows from the above examples (3.17)–(3.24), the main feature of the "golden" representations (3.10) of real numbers in *Bergman's system*, compared to the binary system (3.2), is the *multiplicity* of the "golden" representations of the same real number. The various "golden" representations of one and the same real number can be obtained by using the operations of *convolutions* (3.14) and *devolutions* (3.15) in the "golden" representations (3.10).

A special role among the various "golden" representations (3.10) of one and the same number plays the so-called *minimal form*, which can be obtained from the initial "golden" representation by fulfilling in it all the possible *convolutions* (3.14).

The *minimal form* has the following important features:

1. As the operation of the *convolution* ($011 \rightarrow 100$) is reduced to the transformation of the triple of the neighboring bits **011** into the triple of the neighboring bits **100**, this means that in the *minimal form two bits 1 do not meet alongside.*
2. The *minimal form* has the minimal number of 1's among all the possible "golden" representations of the same number.

### 3.2.8. *Z- and D-properties of natural numbers*

*Bergman's system* (3.1) is a source for new number-theoretical results. The *Z-property* of natural numbers is one of them. This property follows from the next very simple reasoning's.

Let us consider the Φ-*code* (3.9). Note that according to Theorem 3.1, the sum (3.9) is a finite sum for arbitrary natural number $N$.

If we substitute the formula (3.12) instead $\Phi^i$ in the formula (3.9) $N = \sum_i a_i \Phi^i$, we can represent the Φ-*code* (3.9) in the following form:

$$N = \frac{A + B\sqrt{5}}{2}, \quad (3.25)$$

where

$$A = \sum_i a_i L_i \quad (3.26)$$

$$B = \sum_i a_i F_i. \quad (3.27)$$

Note that all the binary numerals in the sums (3.26), (3.27) coincide with the corresponding bits of the Φ-*code* (3.9).

Let us represent now the expression (3.25) as follows:

$$2N = A + B\sqrt{5}. \quad (3.28)$$

Note that the expression (3.28) has general character and is valid for arbitrary natural number $N$.

Let us analyze the expression (3.28). It is clear that the number $2N$, on the left-hand side of (3.28), is always an *even* number. The right-hand side of (3.28) is the sum of number $A$ and the product of number $B$ by the irrational number $\sqrt{5}$. But according to (3.26) and (3.27), $A$ and $B$ are always are integers, because the Fibonacci and Lucas numbers are integers. Then it follows from (3.28) that for the given natural number $N$, the *even* number $2N$ is equal to the sum of integer $A$ and the product of integer $B$ on the irrational number $\sqrt{5}$. And this statement is valid for arbitrary natural number $N$! Then there is the question: when is the identity (3.28) valid for general case? The answer to this question is very simple: the identity (3.28) can be valid for the arbitrary natural number $N$ only for case, if the sum (3.27) is equal to 0 ("zero") and the sum (3.26) is equal to the double number of $N$, that is

$$B = \sum_i a_i F_i \equiv 0 \quad (3.29)$$

$$A = \sum_i a_i L_i \equiv 2N. \quad (3.30)$$

Let us compare now the sums (3.29) and (3.9) $N = \sum_i a_i \Phi^i$. As the bits $a_i$ ($i = 0, \pm 1, \pm 2, \pm 3, \ldots$) in these sums coincide, this means that

(3.29) can be obtained from (3.9) if we replace every power of the golden ratio $\Phi^i$ ($i$ = 0, ±1, ±2, ±3, ...) in the Φ-code (3.9) by the "extended" Fibonacci number $F_i$ ($i$=0,±1,±2,±3, ...). But according to (3.29), the sum (3.29) is equal to 0 independently of the natural number $N$ in (3.28). Thus, we came to new fundamental property of natural numbers, which can be formulated as the following theorem.

**Theorem 3.2. (*Z*-property of natural numbers).** *If we represent an arbitrary natural number N in the Φ-code (3.9) and then substitute the "extended" Fibonacci numbers* $F_i$ *(i = 0, ±1, ±2, ±3, ...) instead of the golden ratio powers* $\Phi^i$ *(i = 0, ±1, ±2, ±3, ...) in the sum (3.9), then the sum that appears as a result of such a substitution is equal to 0, independently of the initial natural number N, that is,*

$$\boxed{\text{For any } N=\sum_i a_i\Phi^i \text{ after substitution } F_i\to\Phi^i \text{ we have}: \sum_i a_iF_i\equiv 0\,(i=0,\pm1,\pm2,\pm3,\ldots)}. \tag{3.31}$$

Compare now the sums (3.30) and (3.9). Because the bits $a_i$ ($i$ = 0, ±1, ±2, ±3, ...) in these sums coincide, this means that the sum (3.30) can be obtained from the sum (3.9), if we replace every power of the golden ratio $\Phi^i$ ($i$ = 0, ±1, ±2, ±3, ...) in the Φ-code (3.9) by the "extended" Lucas number $L_i$ ($i$ = 0, ±1, ±2, ±3, ...). But according to (3.30), the sum (3.30) is equal to $2N$ independently of the initial natural number $N$ in the sum (3.28). Thus, we came to new fundamental property of natural numbers, which can be formulated as the following theorem.

**Theorem 3.3. (*D*-property of natural numbers).** *If we represent an arbitrary natural number N in the Φ-code (3.9) and then substitute the "extended" Lucas numbers* $L_i$ *(i = 0, ±1, ±2, ±3, ...) instead of the golden ratio powers* $\Phi^i$ *(i = 0, ±1, ±2, ±3, ...) in the sum (3.9), then the sum that appears as a result of such a substitution is equal to 2N, independently of the initial natural number N, that is,*

$$\boxed{\text{For any } N=\sum_i a_i\Phi^i \text{ after substitution } L_i\to\Phi^i \text{ we have}: \sum_i a_iL_i\equiv 2N\,(i=0,\pm1,\pm2,\pm3,\ldots)}. \tag{3.32}$$

### 3.2.9. *F- and L-codes*

The above discovered *Z*- and *D*-properties, given by Theorems 3.2 and 3.3, allow new and unusual codes for the representation of natural numbers. Because according to the *Z*-property the sum (3.29) is equal to 0, that is $B = 0$, we can write the expression (3.28) as follows:

$$N = \frac{1}{2}(A + B), \tag{3.33}$$

where $A$ is defined by the expression (3.26) and $B$ by expression (3.27).

By using the expressions (3.26) and (3.27), we can rewrite the expression (3.33) as follows:

$$N = \frac{1}{2}\left(\sum_i a_i L_i + \sum_i a_i F_i\right) = \sum_i a_i \frac{L_i + F_i}{2}. \tag{3.34}$$

Taking into consideration the following well-known identity [53–55]

$$F_{i+1} = \frac{L_i + F_i}{2}, \tag{3.35}$$

we get from (3.34) the following representation of the same natural number $N$:

$$N = \sum_i a_i F_{i+1}. \tag{3.36}$$

The sum (3.36) is called *F-code of natural number N*.

As the bits $a_i$ ($i = 0, \pm1, \pm2, \pm3, \ldots$) in the sums (3.9) and (3.36) coincide, it follows from this fact that the *F*-code of the arbitrary natural number $N$ can be obtained from the Φ-code (3.9) of the same natural number $N$ by the replacement of the golden ratio powers $\Phi^i$ ($i = 0, \pm1, \pm2, \pm3, \ldots$) in the sum (3.9) by the "extended" Fibonacci number $F_{i+1}$ ($i = 0, \pm1, \pm2, \pm3, \ldots$), respectively, that is, $F_{i+1} \to \Phi^i$.

Represent now the *F*-code of $N$ (3.36) as follows:

$$N = \left(\sum_i a_i F_{i+1}\right) + 2B \tag{3.37}$$

where the term $B$ is defined by the sum (3.27) $B=\sum_i a_i F_i$, which is equal to 0 ($B \equiv 0$) according to (3.29). Then the sum (3.37) can be represented as follows:

$$N=\left(\sum_i a_i F_{i+1}\right)+2\left(\sum_i a_i F_i\right)=\sum_i a_i\left(F_{i+1}+2F_i\right). \quad (3.38)$$

Taking into consideration the following well-known formula [53–55]

$$L_{i+1}=F_{i+1}+2F_i, \quad (3.39)$$

the sum (3.38) can be represented as follows:

$$N=\sum_i a_i L_{i+1}. \quad (3.40)$$

The expression (3.40) is called *L-code of natural number N*.

As the bits $a_i$ ($i = 0, \pm 1, \pm 2, \pm 3, \ldots$) in the sums (3.9) and (3.40) coincide, it follows from this fact that the $L$-code of $N$ (3.40) can be obtained from the Φ-code (3.9) of the same natural number $N$ by the replacement of the golden ratio powers $\Phi^i$ ($i = 0, \pm 1, \pm 2, \pm 3, \ldots$) in the sum (3.9) by the "extended" Lucas numbers $L_{i+1}$ ($i = 0, \pm 1, \pm 2, \pm 3, \ldots$), that is, $L_{i+1} \to \Phi^i$.

It is clear that the $L$-code of $N$ (3.40) can also be obtained from the $F$-code of the same number $N$ (3.38) by the replacement of the Fibonacci numbers $F_{i+1}$ in the formula (3.38) by the Lucas numbers $L_{i+1}$, that is, $L_{i+1} \to F_{i+1}$.

Let us represent the sums (3.9), (3.36) and (3.40) in the abridged form (3.10). It is clear that the sums (3.9), (3.36) and (3.40) give three different methods of the binary representation of one and the same natural number $N$. The Φ-code (3.9) is a representation of the natural number $N$ as the sum of the golden ratio powers $\Phi^i$ ($i = 0, \pm 1, \pm 2, \pm 3, \ldots$), the $F$-code (3.36) is a representation of the same natural number $N$ as the sum of the "extended" Fibonacci numbers $\Phi^i$ ($i = 0, \pm 1, \pm 2, \pm 3, \ldots$) and the $L$-code (3.40) is a representation of the same natural number $N$ as the sum of the "extended" Lucas numbers $L_{i+1}$ ($i = 0, \pm 1, \pm 2, \pm 3, \ldots$). As we mentioned above, all the sums (3.9), (3.36) and (3.40), which represent one and the same natural number $N$, have one and the same "golden" representation (3.10).

### 3.2.10. *The left-shift and right-shift of the Φ-, F- and L-codes*

Although the "golden" representations (3.10) of one and the same natural number $N$ coincide for the $\Phi$-, $F$- and $L$-codes, the distinction among the "golden" representations (3.10) of these codes arises, when we perform left-shift or right-shift of the "golden" representations (3.10).

**The left-shift and right-shift of the Φ-code**

Let us denote by $N_{(k)}$ and $N_{(-k)}$ the results of the left-shift and right-shift of the "golden" representation (3.10), respectively. If we interpret the "golden" representation (3.10) as the Φ-code of natural number $N$, then its left-shift (that is, to the side of the higher digits) on the one digit corresponds to the multiplication of the number $N$ by the base Φ, and its right-shift (that is, to the side of the lower digits) on the one digit corresponds to the division of the number $N$ by the base Φ, that is,

$$N_{(1)} = N \times \Phi = \sum_i a_i \Phi^{i+1} \tag{3.41}$$

$$N_{(-1)} = N \times \Phi^{-1} = \sum_i a_i \Phi^{i-1}. \tag{3.42}$$

It is clear that the left-shift of the "golden" representation (3.10) in the Φ-code (3.9) on the $k$ digits corresponds to the multiplication of the number $N$ by $\Phi^k$ and the right-shift on the $k$ digits to the division by $\Phi^k$, that is,

$$N_{(k)} = N \times \Phi^k = \sum_i a_i \Phi^{i+k} \tag{3.43}$$

$$N_{(-k)} = N \times \Phi^{-k} = \sum_i a_i \Phi^{i-k}. \tag{3.44}$$

**The left-shift and right-shift of the F-code**

Let us consider the left-shift and right-shift of the "golden" representation (3.10), when we interpret it in the $F$- or $L$-codes. If we interpret the "golden" representation (3.10) in the $F$-code, then its left-shift on the $k$ digits leads us to the following expression:

$$N_{(k)} = \sum_i a_i F_{i+1+k}. \tag{3.45}$$

Apply to the expression (3.45) the following well-known identity for the generalized Fibonacci numbers $G_k = G_{k-1} + G_{k-2}$ [53–55]:

$$G_{n+m} = F_{m-1}\,G_n + F_m\,G_{n+1}. \tag{3.46}$$

For the case $G_k = F_k$ the identity (3.46) takes the following form:

$$F_{n+m} = F_{m-1}\,F_n + F_m\,F_{n+1}. \tag{3.47}$$

It follows from (3.47) that for the cases $n = i$ and $m = k + 1$ the identity (3.47) is reduced to

$$F_{i+1+k} = F_k F_i + F_{k+1} F_{i+1}. \tag{3.48}$$

By substituting (3.48) into the sum (3.45), we get:

$$N_{(k)} = \sum_i a_i F_{i+1+k} = \sum_i a_i \left(F_k F_i + F_{k+1} F_{i+1}\right) = F_k \sum_i a_i F_i + F_{k+1} \sum_i a_i F_{i+1}. \tag{3.49}$$

Taking into consideration the $Z$-property (3.29) $B = \sum_i a_i F_i \equiv 0$ and the sum (3.36) $N = \sum_i a_i F_{i+1}$ for the $F$-code, we can simplify the expression (3.49) as follows:

$$N_{(k)} = \sum_i a_i F_{i+1+k} = F_{k+1} \times N. \tag{3.50}$$

If we interpret the "golden" representation (3.10) in the $F$-code, then its right-shift on the $k$ digits leads us to the following sum:

$$N_{(-k)} = \sum_i a_i F_{i+1-k}. \tag{3.51}$$

Let us consider now the identity (3.47) $F_{n+m} = F_{m-1}\,F_n + F_m\,F_{n+1}$. If we take $n = i$ and $m = -k + 1$, then we can write the identity (3.47) as follows:

$$F_{i+1-k} = F_{-k} F_i + F_{-k+1} F_{i+1}. \tag{3.52}$$

By substituting (3.52) into the sum (3.50), after the simple transformations with regard to (3.29) $B=\sum_i a_i F_i \equiv 0$ (*Z*-property) and (3.36) $N=\sum_i a_i F_{i+1}$ (*F*-code), we get:

$$N_{(-k)} = \sum_i a_i F_{i+1-k} = F_{-k+1} \times N. \quad (3.53)$$

Let us formulate now the results (3.50), (3.53) in the following theorem.

**Theorem 3.4.** *The left-shift of the "golden" representation (3.10), interpreted as the F-code, on the k digits, corresponds to the multiplication of the number N by the "extended" Fibonacci number* $F_{k+1}$, *but its right-shift on the k digits corresponds to the multiplication of the number N by the "extended" Fibonacci number* $F_{-k+1}$.

Let us consider now the formula (3.53). For the case $k = 1$ (the right-shift on the one digit) the formula (3.53) takes the following form:

$$\sum_i a_i F_i = F_0 \times N. \quad (3.54)$$

But the Fibonacci number $F_0 = 0$ and therefore the formula (3.54) is reduced to the formula (3.29) $B=\sum_i a_i F_i \equiv 0$ (*Z*-property). This consideration is another proof of the *Z*-property, given by Theorem 3.2.

Note that the right-shift of the "golden" representation (3.10), interpreted in the *F*-code, on the three digits corresponds to the multiplication of the number $N$ by the Fibonacci number $F_{-2} = -1$. This means that such right-shift leads us to the number $(-N) = (-1) \times N$. This property of the *F*-code together with the *Z*-property $B=\sum_i a_i F_i \equiv 0$ lead us to number of useful applications in computer science and digital metrology.

***The left-shift and right-shift of the L-code***

Let us consider now the left-shift and right-shift of the "golden" representation (3.10), interpreted as the $L$-code (3.40). Its left-shift and right-shift on the $k$ digits lead to the following sums, respectively:

$$N_{(k)} = \sum_i a_i L_{i+1+k} \tag{3.55}$$

$$N_{(-k)} = \sum_i a_i L_{i+1-k}. \tag{3.56}$$

By using the identity (3.46) $G_{n+m} = F_{m-1}G_n + F_m G_{n+1}$ and taking $G_k = L_k$, we can express the "extended" Lucas numbers $L_{i+1+k}$ and $L_{i+1-k}$ as follows:

$$L_{i+1+k} = F_i L_k + F_{i+1} L_{k+1} \tag{3.57}$$

$$L_{i+1-k} = F_i L_{-k} + F_{i+1} L_{-k+1}. \tag{3.58}$$

Then the expressions (3.55), (3.56) can be represented as follows, respectively:

$$N_{(k)} = \sum_i a_i L_{i+1+k} = L_k \sum_i a_i F_i + L_{k+1} \sum_i a_i F_{i+1} \tag{3.59}$$

$$N_{(-k)} = \sum_i a_i L_{i+1-k} = L_{-k} \sum_i a_i F_i + L_{-k+1} \sum_i a_i F_{i+1}. \tag{3.60}$$

With regard to the expressions (3.29) $B = \sum_i a_i F_i \equiv 0$ ($Z$-property) and (3.36) $N = \sum_i a_i F_{i+1}$ ($F$-code), we get from (3.59), (3.60) the following results:

$$N_{(k)} = \sum_i a_i L_{i+1+k} = L_{k+1} \times N \tag{3.61}$$

$$N_{(-k)} = \sum_i a_i L_{i+1-k} = L_{-k+1} \times N. \tag{3.62}$$

We can formulate the results (3.61), (3.62) as follows.

**Theorem 3.5.** *The left-shift of the "golden" representation (3.10), interpreted as the L-code, corresponds to the multiplication of the* *number N by the "extended" Lucas number $L_{k+1}$, but its right-shift on the k digits corresponds to the multiplication of the number N by the "extended" Lucas number $L_{-k+1}$.*

Let us consider the formula (3.62). For the case $k = 1$ (the right-shift on the one digit) the formula (3.62) takes the following form:

$$\sum_i a_i L_i = L_0 \times N. \tag{3.63}$$

As the "extended" Lucas number $L_0 = 2$, the formula (3.63) is reduced to the formula (3.30) $A = \sum_i a_i L_i \equiv 2N$ (*D*-property). This consideration is another proof of the *D*-property, given by Theorem 3.3.

### 3.2.11. *Numerical examples*

Consider again the "golden" representation (3.10). We can see that the "golden" representation (3.10) is separated by the point into two parts, namely the left part, which consists of the digits with the non-negative indices, and the right part, which consists of the digits with the negative indices. As example we can consider the "golden" representation of the decimal number 10 in *Bergman's system*:

$$10_{10} = 10100.0101. \tag{3.64}$$

For the Φ-code (3.9), the "golden" representation (3.64) has the following algebraic interpretation:

$$10_{10} = \Phi^4 + \Phi^2 + \Phi^{-2} + \Phi^{-4}. \tag{3.65}$$

By using the formula (3.12) $\Phi^i = \frac{L_i + F_i\sqrt{5}}{2} \left(i = 0, \pm1, \pm2, \pm3, \ldots\right)$, we can represent the sum (3.64) as follows:

$$10_{10} = \Phi^4 + \Phi^2 + \Phi^{-2} + \Phi^{-4} = \frac{L_4 + F_4\sqrt{5}}{2} + \frac{L_2 + F_2\sqrt{5}}{2} + \frac{L_{-2} + F_{-2}\sqrt{5}}{2} + \frac{L_{-4} + F_{-4}\sqrt{5}}{2}. \tag{3.66}$$

If we take into consideration the following correlations for the "extended" Fibonacci and Lucas numbers:

$$L_{-2} = L_2;\ L_{-4} = L_4;\ F_{-2} = -F_2;\ F_{-4} = -F_4,$$

then we get from the expression (3.65) the following result:

$$10 = \frac{2(L_4 + L_2)}{2} = L_4 + L_2 = 7 + 3.$$

Let us consider now the interpretation of the "golden" representation (3.10) as the $F$- and $L$-codes:

$$10_{10} = F_5 + F_3 + F_{-1} + F_{-3} = 5 + 2 + 1 + 2;$$

$$10_{10} = L_5 + L_3 + L_{-1} + L_{-3} = 11 + 4 - 1 - 4.$$

Also we can check the sum (3.64) by the $Z$- and $D$-properties. If we change in the formula (3.64) $10_{10} = \Phi^4 + \Phi^2 + \Phi^{-2} + \Phi^{-4}$ all the powers $\Phi^i$ by the "extended" Fibonacci ($F_i$) and Lucas ($L_i$) numbers, we get the following sums:

$$F_4 + F_2 + F_{-2} + F_{-4} = 3 + 1 + (-1) + (-3) = 0 \ (Z\text{-}property)$$

$$L_4 + L_2 + L_{-2} + L_{-4} = 7 + 3 + 3 + 7 = 20 = 2 \times 10 \ (D\text{-}property).$$

### 3.2.12. *Algebraic summation of integers*

We can use the $Z$-property to check arithmetical operations. For example, consider the operation of the algebraic summation of two integers $N_1 \pm N_2$. As an outcome of this operation we always obtain new integer. This means that the "golden" algebraic summation of two integers, represented in the $\Phi$-, $F$- or $L$-codes, always results in the new "golden" representation of the algebraic sum $N_1 \pm N_2$ in the $\Phi$-, $F$- or $L$-codes. It follows from this consideration that *the Z-property is invariant about the "golden" algebraic summation.* Similar conclusion is valid for the *"golden" multiplication.* As the outcome of the "golden" division of two integers is always two integers, the quotient $Q$ and the remainder $R$, it follows from this consideration that *the results of the "golden" division, the integers Q and R, preserve the Z-property.*

Hence, we have obtained some new fundamental properties of natural numbers, represented in the $\Phi$-, $F$- and $L$-codes. These properties (for example the $Z$-property) are invariant about arithmetical operations and may be used for checking arithmetical operations in computers.

In conclusion, we note that Theorems 3.1–3.5 are true only for natural numbers. This means that Theorems 3.1–3.5 express new fundamental properties of natural numbers. It is surprising to many mathematicians to know that the new mathematical properties of natural numbers, given in Theorems 3.1–3.5, were only discovered at the beginning of the 21st century, that is, 2500 years after the beginning of the theoretical study of natural numbers. The golden ratio and the "extended" Fibonacci and Lucas numbers play a fundamental role in this discovery. This discovery connects together two outstanding mathematical concepts of Greek mathematics — *natural numbers* and *golden ratio*. This discovery is the confirmation of fundamental role of *Bergman's system* (3.1) for the development of the "golden" number theory, described in [33].

However, the ternary "golden" mirror-symmetrical arithmetic, described by the author in article [28], is the most interesting application of *Bergman's system* in computer science.

## 3.3. The "golden" ternary mirror-symmetrical representation

### 3.3.1. *Conversion of the binary "golden" representation to the ternary "golden" representation*

We start from the Φ-code of natural number (3.9) $N = \sum_i a_i \Phi^i$, in which we use only the *minimal forms* to represent numbers. We recall the following fundamental property of the MINIMAL FORM:

| $a_{k+1}$ | $a_k$ | $a_{k-1}$ |
|---|---|---|
| **0** | 1 | **0** |

. (3.67)

This means that in the *minimal form* each bit $a_k = 1$ is "surrounded" by the two adjacent bits $a_{k+1} = a_{k-1} = 0$ (in bold).

Let us consider now the following fundamental identity for the powers of the golden ratio:

$$\Phi^{k+1} = \Phi^k + \Phi^{k-1} \, (k = 0, \pm1, \pm2, \pm3, \ldots). \quad (3.68)$$

The identity (3.68) can be rewritten as follows:

$$\Phi^k = \Phi^{k+1} - \Phi^{k-1} \; (k = 0, \pm 1, \pm 2, \pm 3, \ldots). \tag{3.69}$$

By using (3.69), we can perform the following transformation over (3.67):

| | $a_{k+1}$ | $a_k$ | $a_{k-1}$ | $\Rightarrow$ | $a_{k+1}$ | $a_k$ | $a_{k-1}$ |
|---|---|---|---|---|---|---|---|
| $\Phi^k =$ | **0** | 1 | **0** | $\Rightarrow$ | **1** | 0 | $\bar{\mathbf{1}}$ |

, (3.70)

where $\bar{1}$ is the negative 1, that is, $\bar{1} = -1$. It follows from (3.70) that the positive binary 1 of the $k$th digit $a_k = 1$ can be transformed into two 1's, the positive numeral of 1 of the $(k + 1)$th digit $a_{k+1} = 1$ and the negative numeral of $\bar{1}$ of the $(k - 1)$th digit $a_{k-1} = \bar{1}$.

The transformation (3.70) can be used for the conversion of the *minimal form* of the *binary "golden" representation* (3.10) of the natural number $N$ into the *ternary "golden" representation* of the same integer $N$.

Let us consider now the binary "golden" representation (3.10) of the number $N = 5$, represented in the *minimal form*:

| $k$ | 4 | 3 | 2 | 1 | 0 | −1 | −2 | −3 | −4 |
|---|---|---|---|---|---|---|---|---|---|
| $N = 5$ | 0 | **1** | 0 | 0 | 0. | **1** | 0 | 0 | 1 |

. (3.71)

Let us convert the binary "golden" representation (3.71) of the number $N = 5$, represented in the *minimal form*, into the ternary "golden" representation of the same number $N = 5$. With this purpose, we can apply the code transformation (3.70) simultaneously to all digits, being the binary 1's and having the odd indices ($k = 2m + 1$). We can see that the transformation (3.70) can be applied for the example (3.71) only for the 3rd and (−1)th digits, which are the binary 1's (in bold). As a result of such transformations, we get the following ternary "golden" representation of the natural number $N = 5$:

| $k$ | 4 | 3 | 2 | 1 | 0 | −1 | −2 | −3 | −4 |
|---|---|---|---|---|---|---|---|---|---|
| $N = 5$ | **1** | 0 | $\bar{\mathbf{1}}$ | 0 | **1.** | 0 | $\bar{\mathbf{1}}$ | 0 | 1 |

. (3.72)

We can see from (3.72) that all digits with the odd indices $k = 2m + 1$ (3, 1, −1, −3) are equal to 0 identically, but the digits with the even indices $k = 2m$ (4, 2, −2, −4) take the ternary values from the set $\{\bar{1}, 0, 1\}$. This means that all digits with the *odd* indices $k = 2m + 1$ (3, 1, −1, −3) are "non-informative," because their values are equal to 0 identically and they do not influence on the value of the number $N = 5$. Omitting in (3.72) all the "non-informative" digits, we get the following ternary "golden" representation of the initial natural number $N = 5$:

| $k$ | 4 | 2 | 0 | −2 | −4 |
|---|---|---|---|---|---|
| $N = 5$ | 1 | $\bar{1}$ | 1. | $\bar{1}$ | 1 |

. (3.73)

Algebraic interpretation of the ternary "golden" representation (3.73) has the form of the following sum:

$$N = 5 = 1\times\Phi^4 + \bar{1}\times\Phi^2 + 1\times\Phi^0 + \bar{1}\times\Phi^{-2} + 1\times\Phi^{-4}$$

$$= \Phi^4 - \Phi^2 + \Phi^0 - \Phi^{-2} + \Phi^{-4}. \quad (3.74)$$

In general, if we convert the binary "golden" representation (3.10), represented in the *minimal form*, into the ternary "golden" representation, then after omitting the "non-informative" digits we get the following sum:

$$N = \sum_i a_{2i}\Phi^{2i}, \quad (3.75)$$

where $a_{2i}$ is the ternary numeral of the $(2i)$th digit.

We can perform the following digit enumeration for the ternary numeral system (3.75). Each ternary digit $a_{2i}$ is replaced by the ternary digit $c_i$: $a_{2i} \to c_i \ (i = 0, \pm1, \pm2, \pm3, \ldots)$. As a result of such enumeration, we get the expression (3.75) in the following form:

$$N = \sum_i c_i\Phi^{2i}, \quad (3.76)$$

where $c_i \in \{\bar{1}, 0, 1\}$ is the ternary numeral of the *i*th digit; $\Phi^{2i}$ is the weight of the *i*th digit; $\Phi^2$ is the *base* of the numeral system (3.76). We name the sum (3.76) the *ternary* $\Phi$*-code* of natural number $N$.

With regard to the expression (3.76), the ternary "golden" representation (3.73) of the natural number $N = 5$ takes the following form:

| $k$ | 2 | 1 | 0 | −1 | −2 |
|---|---|---|---|---|---|
| $N = 5$ | 1 | $\bar{1}$ | 1. | $\bar{1}$ | 1 |

. (3.77)

The conversion of the binary Φ-code (3.9) of the natural number $N$ into the ternary Φ-code (3.76) of the same natural number $N$ can be fulfilled by means of a simple combinative logic circuit, which transforms the adjacent three binary digits $a_{2i+1}\, a_{2i}\, a_{2i-1}$ of the initial binary "golden" *minimal form* into the ternary informative digit $c_i$ of the ternary "golden" representation in accordance with Table 3.2.

Note that Table 3.2 uses only the 5 binary code combinations from the 8 possible 3-bit binary code combinations, because the initial binary "golden" representation of the kind (3.71) is represented in the *minimal form* and the binary code combinations {011, 110, 111} are *prohibited* for this situation.

Table 3.2. Conversion of the binary digits $a_{2i+1}\, a_{2i}\, a_{2i-1}$ into the ternary digit $c_i$.

| $a_{2i+1}$ | $a_{2i}$ | $a_{2i-1}$ | $c_i$ |
|---|---|---|---|
| 0 | 0 | 0 | 0 |
| 0 | 0 | 1 | 1 |
| 0 | 1 | 0 | 1 |
| 1 | 0 | 0 | $\bar{1}$ |
| 1 | 0 | 1 | 0 |

The code transformations, given with the 2nd and 4th rows of Table 3.2: 000 → 0 and 010 → 1, respectively, are trivial. The code transformations, given with the 3rd, 5th and 6th rows of Table 3.2 $\left(001 \to 1, 100 \to \bar{1}, 101 \to 0\right)$, follow directly from (3.70). For instance, the code transformation of the 6th row 101 → 0 means that the negative numeral ($\bar{1}$), arising in accordance with (3.70) from the left binary digit $a_{2i+1} = 1$, is summarized with the positive numeral 1, arising from the right binary digit $a_{2i+1}$ = 1. It follows from this consideration that their sum $\left(\bar{1}+1\right)$ equals to the ternary numeral $c_i = 0$.

### 3.3.2. *The ternary "golden" F- and L-representations*

In the above we have introduced the so-called $F$- and $L$-codes (3.36) and (3.40) for *Bergman's system* (3.9). Recall that these unusual codes are equivalent to the Φ-code (3.9) of the same natural number $N$. By using the ternary Φ-code of natural number $N$, given by (3.76), it is easy to write the ternary $F$- and $L$-codes of the same natural number $N$ in the following forms:

$$N = \sum_i c_i F_{2i+1}, \tag{3.78}$$

$$N = \sum_i c_i L_{2i+1}. \tag{3.79}$$

Note that the values of the ternary digits in (3.76), (3.77), (3.79) coincide. It follows from this consideration that the ternary "golden" Φ-, $F$-, $L$-representations of the number $N = 5$, given by the example (3.77), have three different algebraic interpretations:

(a) The ternary Φ-code:

$$\begin{aligned}5 &= 1\times\Phi^4 + \bar{1}\times\Phi^2 + 1\times\Phi^0 + \bar{1}\times\Phi^{-2} + 1\times\Phi^{-4} = \Phi^4 - \Phi^2 + \Phi^0 - \Phi^{-2} + \Phi^{-4}\\ &= \frac{L_4 + F_4\sqrt{5}}{2} - \frac{L_2 + F_2\sqrt{5}}{2} + 1 - \frac{L_{-2} + F_{-2}\sqrt{5}}{2} + \frac{L_{-4} + F_{-4}\sqrt{5}}{2}\\ &= L_4 - L_2 + 1 = 7 - 3 + 1.\end{aligned}$$

(b) The ternary $F$-code:

$$5 = 1\times F_5 + \bar{1}\times F_3 + 1\times F_1 + \bar{1}\times F_{-1} + 1\times F_{-3} = 5 - 2 + 1 - 1 + 2.$$

(c) The ternary $L$-code:

$$5 = 1\times L_5 + \bar{1}\times L_3 + 1\times L_1 + \bar{1}\times L_{-1} + 1\times L_{-3} = 11 - 4 + 1 + 1 - 4.$$

Note that we use for the calculation of the above ternary "golden" Φ-, $F$- and $L$-representations the properties (3.8) $F_{-n} = (-1)^{n+1}F_n$; $L_{-n} = (-1)^n L_n$ and Table 3.1:

| $n$ | 0 | 1 | 2 | 3 | 4 | 5 | 6 | 7 | 8 | 9 | 10 |
|---|---|---|---|---|---|---|---|---|---|---|---|
| $F_n$ | 0 | 1 | 1 | 2 | 3 | 5 | 8 | 13 | 21 | 34 | 55 |
| $F_{-n}$ | 0 | 1 | −1 | 2 | −3 | 5 | −8 | 13 | −21 | 34 | −55 |
| $L_n$ | 2 | 1 | 3 | 4 | 7 | 11 | 18 | 29 | 47 | 76 | 123 |
| $L_{-n}$ | 2 | −1 | 3 | −4 | 7 | −11 | 18 | −29 | 47 | −76 | 123 |

### 3.3.3. *The representation of negative numbers*

Similarly to the ternary-symmetrical numeral system (2.8) $N=\sum_{i=0}^{n-1} b_i 3^i$, the important advantage of the ternary numeral system (3.76) is a possibility to represent both positive and negative numbers in the "direct" code. The ternary "golden" representation of the negative number ($-N$) can be obtained from the ternary "golden" Φ-representation of the initial natural number $N$ by means of the application of the "ternary inversion" rule (2.17) $\bar{1}\to 1;\ 0\to 0;\ 1\to\bar{1}$. Applying this rule to the ternary "golden" Φ-representation (3.77) of number 5, we get the following ternary "golden" Φ-representation of the negative number (−5):

| $k$ | 2 | 1 | 0 | −1 | −2 |
|---|---|---|---|---|---|
| $N=-5$ | $\bar{1}$ | 1 | $\bar{1}.$ | 1 | $\bar{1}$ |

. (3.80)

### 3.3.4. *Mirror-symmetrical property of the ternary "golden" representations*

By considering the ternary "golden" Φ-representation (3.76) of number $N$ = 5, we can see that the left part $(1\bar{1})$ of the ternary "golden" Φ-representation (3.77) is mirror-symmetrical to its right part $(\bar{1}1)$ relatively to the 0th digit. This property, named *mirror-symmetrical property* of the numeral system (3.76) [28], is a fundamental property of the *ternary "golden" Φ-, F- and L-representations* of integers (positive and negative). Table 3.3 demonstrates this property for some initial natural numbers.

Let us give the explanations of Table 3.3. The first row $i$ means the digit indices of the 7-digit ternary mirror-symmetrical code (3.76), the second row $\Phi^{2i}$ means the digit weights of the 7-digit ternary mirror-symmetrical $\Phi$-code (3.76), the third row $F_{2i+1}$ means the digit weights of the 7-digit ternary mirror-symmetrical $F$-code (3.78), the fourth row $L_{2i+1}$ means the digit weights of the 7-digit ternary mirror-symmetrical $L$-code (3.79). The fifth row $N$ means positive integers from 0 to 10; their ternary "golden" mirror-symmetrical representations are represented in the rows below the fifth row. All data, relating to the 0th digit, which separates the left and right parts of the ternary "golden" mirror-symmetrical representations of positive integers (see the column 0) are in bold.

Thus, thanks to this simple observation, we have found the most important fundamental property of integers called *mirror-symmetrical property of integers*. Based on this fundamental property, the "ternary numeral system," given by (3.76), was named *ternary mirror-symmetrical numeral system* [28].

Table 3.3. Property of "mirror symmetry".

| $i$ | 3 | 2 | 1 | **0** | −1 | −2 | −3 |
|---|---|---|---|---|---|---|---|
| $\Phi^{2i}$ | $\Phi^6$ | $\Phi^4$ | $\Phi^2$ | $\boldsymbol{\Phi^0}$ | $\Phi^{-2}$ | $\Phi^{-4}$ | $\Phi^{-6}$ |
| $F_{2i+1}$ | 13 | 5 | 2 | **1** | 1 | 2 | 5 |
| $L_{2i+1}$ | 29 | 11 | 4 | **1** | −1 | −4 | −11 |
| $N\downarrow$ | ↓ | ↓ | ↓ | ↓ | ↓ | ↓ | ↓ |
| 0 | 0 | 0 | 0 | **0.** | 0 | 0 | 0 |
| 1 | 0 | 0 | 0 | **1.** | 0 | 0 | 0 |
| 2 | 0 | 0 | 1 | $\bar{\mathbf{1}}$**.** | 1 | 0 | 0 |
| 3 | 0 | 0 | 1 | **0.** | 1 | 0 | 0 |
| 4 | 0 | 0 | 1 | **1.** | 1 | 0 | 0 |
| 5 | 0 | 1 | $\bar{1}$ | **1.** | $\bar{1}$ | 1 | 0 |
| 6 | 0 | 1 | 0 | $\bar{\mathbf{1}}$**.** | 0 | 1 | 0 |
| 7 | 0 | 1 | 0 | **0.** | 0 | 1 | 0 |
| 8 | 0 | 1 | 0 | **1.** | 0 | 1 | 0 |
| 9 | 0 | 1 | 1 | $\bar{\mathbf{1}}$**.** | 1 | 1 | 0 |
| 10 | 0 | 1 | 1 | **0.** | 1 | 1 | 0 |

Another interesting feature of the ternary mirror-symmetric system (3.76) follows from Table 3.3. For all the canonical numeral systems (2.2) $x=\sum_{i=0}^{n-1} b_i R^i$ the "extension" of the positional representation of the number is carried out only to the side of the higher digits. For the ternary mirror-symmetrical system (3.76), the "extension" of the ternary mirror-symmetrical representation occur to both sides, i.e., to the sides of the higher and junior digits simultaneously. This feature, as also the property of "mirror-symmetry" and other features, single out the ternary mirror-symmetrical positional numeral system (3.76) among all other positional numeral systems.

### 3.3.5. *The base of the ternary mirror-symmetrical numeral system*

It follows from (3.76) that the base of the numeral system (3.76) is the square of the golden ratio, that is,

$$\Phi^2 = \frac{3+\sqrt{5}}{2} \approx 2.618. \tag{3.81}$$

This means that the numeral system (3.76) is the numeral system with irrational base.

The base (3.81) has the following traditional representation:

$$\Phi^2 = 10.$$

### 3.3.6. *Comparison of ternary mirror-symmetrical numbers*

Let us consider the set of the weights of the $(2n + 1)$th ternary mirror-symmetrical Φ-code (3.76):

$$\left\{\Phi^{2n}, \Phi^{2(n-1)}, \ldots, \Phi^4, \Phi^2, \Phi^0, \Phi^{-2}, \Phi^{-4}, \ldots, \Phi^{-2(n-1)}, \Phi^{-2n}\right\}. \tag{3.82}$$

It is easy to prove that the weight of the $n$th digit of the ternary mirror-symmetrical system (3.76) is always strictly more than the sum of the rest weights of (3.76) at the right of the $n$th digit. It follows from this fact that the higher significant digit of the ternary mirror-symmetrical Φ-code (3.76) contains the information about the sign of the ternary mirror-symmetrical number. If the numeral of the higher significant digit of the ternary mirror-symmetrical Φ-code (3.76) is equal to 1, this means that the ternary mirror-symmetrical number is positive. If the numeral of the higher significant digit of the ternary mirror-symmetrical Φ-code (3.76) is equal to $\bar{1}$, this means that the ternary mirror-symmetrical number is negative.

From this consideration, there is a very simple method to compare two ternary mirror-symmetrical numbers $A$ and $B$. The comparison begins from the higher digits of the numbers and lasts until we obtain the first pair of the not coincident ternary digits $a_k$ and $b_k$. If the numeral $a_k > b_k \left(1 > 0, 1 > \bar{1}, 0 > \bar{1}\right)$, then $A > B$. In the opposite case: $A < B$.

Hence, we have found two important advantages of the ternary mirror-symmetrical numeral system (3.76):

1. Similarly to the classic ternary-symmetrical numeral system (2.8) $N = \sum_{i=0}^{n-1} b_i 3^i$, the sign of the ternary mirror-symmetrical number is determined by the higher significant $\left(1 \text{ or } \bar{1}\right)$ digit of the ternary mirror-symmetrical numeral system (3.76).
2. Comparison of the numbers is similar to the classic ternary-symmetrical numeral system (2.8), that is, from the higher digits to the right until we obtain the first pair of the not coincident ternary digits.

### 3.3.7. *The range of the ternary mirror-symmetrical representation of numbers*

Let us consider the range of number representation in the numeral system (3.76). Suppose that the ternary Φ-code (3.76) has $2m + 1$ ternary digits. In this case, by using (3.76) we can represent all integers (positive and negative) in the range from

$$N_{\max} = \underbrace{1\,1\,\ldots\,1}_{m}\,1.\,\underbrace{1\,1\,\ldots\,1}_{m} \tag{3.83}$$

to

$$N_{\min} = \underbrace{\bar{1}\,\bar{1}\,\ldots\,\bar{1}}_{m}\,\bar{1}.\,\underbrace{\bar{1}\,\bar{1}\,\ldots\,\bar{1}}_{m}. \tag{3.84}$$

It is clear that $N_{max}$ is a positive number and $N_{min}$ is a negative number.

It is clear that $N_{min}$ is the ternary inversion of $N_{max}$, that is, they are equal by absolute value:

$$|N_{min}| = N_{max}. \tag{3.85}$$

It follows from this consideration that by using the $2m + 1$ ternary Φ-code (3.76), we can represent

$$2N_{max} + 1 \tag{3.86}$$

integers (positive and negative), by including the number 0.

For the calculation of $N_{max}$ we can interpret (3.83) as the ternary $L$-code (3.79). Taking into consideration the expression (3.79) for the $L$-code, we can give the following algebraic interpretation of (3.83):

$$N_{max} = L_{2m+1} + L_{2m-1} + \ldots + L_5 + L_3 + L_1 + L_{-1} + L_{-3} + L_{-5} + \ldots + L_{-(2m-1)}. \tag{3.87}$$

As it follows from Table 1.4 (see Chapter 1) for the "extended" Lucas numbers

| $n$ | 0 | 1 | 2 | 3 | 4 | 5 | 6 | 7 | 8 | 9 | 10 |
|---|---|---|---|---|---|---|---|---|---|---|---|
| $L_n$ | 2 | 1 | 3 | 4 | 7 | 11 | 18 | 29 | 47 | 76 | 123 |
| $L_{-n}$ | 2 | −1 | 3 | −4 | 7 | −11 | 18 | −29 | 47 | −76 | 123 |

,

we have the following simple relation between the Lucas numbers $L_{-i}$ and $L_i$:

$$L_{-i} = (-1)^i L_i \; (i = 0, \pm1, \pm2, \pm3, \ldots). \tag{3.88}$$

This means that for the *odd* indices $i = 1, 3, 5, \ldots, 2k - 1$ we have the following property for Lucas numbers [28]:

$$L_{-(2k-1)} = -L_{2k-1} \; (k = 1, 2, 3, \ldots). \tag{3.89}$$

Taking into consideration the property (3.89), we can rewrite the sum (3.87) as follows:

$$N_{max} = L_{2m+1} + L_{2m-1} + \ldots + L_5 + L_3 + L_1 - L_1 - L_3 - L_5 - \ldots - L_{2m-1}, \tag{3.90}$$

from which we get the following result:

$$N_{max} = L_{2m+1}. \tag{3.91}$$

Taking into consideration (3.86) and (3.91), we can formulate the following theorem.

**Theorem 3.6.** *By using* $(2m+1)$ *ternary digits* $\{\bar{1},0,1\}$, *we can represent in the ternary mirror-symmetrical numeral system (3.76) the* $2L_{2m+1}+1$ *integers (including the* $L_{2m+1}$ *positive integers, the* $L_{2m+1}$ *negative integers and the number of 0) in the range from* $(-L_{2m+1})$ *to* $(+L_{2m+1})$, *where* $L_{2m+1}$ *is the Lucas number.*

### 3.3.8. *Code redundancy of the ternary mirror-symmetrical numeral system*

The $(2m+1)$-digit ternary Φ-code (3.76) $N=\sum_i c_i\Phi^{2i}$ is redundant ternary numeral system. For the calculation of the *code redundancy* of the $(2m+1)$-digit ternary Φ-code (3.76) we compare the ranges of number representations for the $(2m+1)$-digit ternary Φ-code (3.76) and the ternary-symmetrical numeral system (2.8) $N=\sum_{i=0}^{n-1} b_i 3^i$, which is non-redundant ternary numeral system.

For the calculation of the relative redundancy $R$, we will use the following formula:

$$R=\frac{k-n}{k}=1-\frac{n}{k}, \tag{3.92}$$

where $k$ and $n$ are the digit numbers of the redundant ternary mirror-symmetrical numeral system (3.76) $N=\sum_i c_i\Phi^{2i}$ and non-redundant ternary-symmetrical numeral systems (2.8) $N=\sum_{i=0}^{n-1} b_i 3^i$ for the representation of the same range of numbers, respectively.

According to Theorem 3.6, the $(2m+1)$-digit ternary mirror-symmetrical Φ-code (3.76) can represent $(2L_{2m+1}+1\approx 2L_{2m+1})$ integers in the range from $-L_{2m+1}$ to $L_{2m+1}$, where $L_{2m+1}$ is the Lucas number.

According to Theorem 2.1 (see Chapter 2), the non-redundant $n$-digit ternary-symmetrical code (2.8) $N=\sum_{i=0}^{n-1} b_i 3^i$ can represent $3^n$ integers in the range from $N_{\min}=-\frac{3^n-1}{2}$ up to $N_{\max}=\frac{3^n-1}{2}$. For the calculation of the code redundancy of the ternary mirror-symmetrical Φ-code (3.76), we have to equate the ranges of number representation of these two ternary codes, that is

$$3^n \approx 2L_{2m+1} \tag{3.93}$$

and then calculate the number of digit $n$, necessary for the representation of the range $2L_{2m+1}$. To solve this task, we express the Lucas number $L_{2m+1}$ through the golden ratio $\Phi=\frac{1+\sqrt{5}}{2}$ by using Binet's formula for Lucas numbers. By using Binet's formula (1.51) $L_n=\begin{cases}\Phi^n+\Phi^{-n} \text{ for } n=2k\\ \Phi^n-\Phi^{-n} \text{ for } n=2k+1\end{cases}$, we can write the following approximate formula for the calculation of the Lucas number $L_{2m+1}$:

$$L_{2m+1} \approx \Phi^{2m+1}. \tag{3.94}$$

After substituting (3.94) into (3.93), we get the following approximate formula for the comparison of the ranges:

$$3^n \approx 2\Phi^{2m+1} \tag{3.95}$$

If we take the logarithm with base 3 from both parts of the equality (3.95), we get the following formula for the calculation of $n$:

$$n=(2m+1)\log_3 \Phi+\log_3 2 \approx (2m+1)\log_3 \Phi. \tag{3.96}$$

The expression (3.96) can be used for the calculation of the code redundancy of the ternary mirror-symmetrical Φ-code (3.76). For this purpose we substitute into the formula (3.92) the following expressions instead $k$ and $n$:

$$k=2m+1 \text{ and } n=(2m+1)\log_3 \Phi. \tag{3.97}$$

As a result, we get the following value for the relative redundancy of the ternary mirror-symmetrical Φ-code (3.77):

$$R = 1 - \frac{n}{k} = 1 - \frac{(2m+1)\log_3 \Phi}{(2m+1)} = 1 - \log_3 \Phi \approx 0.562(56.2\%). \quad (3.98)$$

## 3.4. The ternary mirror-symmetrical summation and subtraction

### 3.4.1. *Mirror-symmetrical summation-subtraction*

The following identities for the golden ratio powers underlie the mirror-symmetric summation-subtraction:

$$2\Phi^{2k} = \Phi^{2(k+1)} - \Phi^{2k} + \Phi^{2(k-1)} \quad (3.99)$$

$$3\Phi^{2k} = \Phi^{2(k+1)} + 0 + \Phi^{2(k-1)} \quad (3.100)$$

$$4\Phi^{2k} = \Phi^{2(k+1)} + \Phi^{2k} + \Phi^{2(k-1)}, \quad (3.101)$$

where $k = 0, \pm1, \pm2, \pm3, \ldots$ .

The identity (3.99) underlies the mirror-symmetrical summation-subtraction of two single-digit ternary digits and gives the rule of the carry-over and intermediate sum formation (Table 3.4).

Table 3.4. Mirror-symmetrical summation (subtraction).

| $a_k + b_k$ | $\bar{1}$ | 0 | 1 |
|---|---|---|---|
| $\bar{1}$ | $\bar{1}1\bar{1}$ | $\bar{1}$ | 0 |
| 0 | $\bar{1}$ | 0 | 1 |
| 1 | 0 | 1 | $1\bar{1}1$ |

The main peculiarity of Table 3.4 appears at the summation (subtraction) of the two ternary numerals $a_k + b_k$ with equal signs, i.e.

$$\begin{array}{ccccccc} a_k & + & b_k & = & c_k & s_k & c_k \\ 1 & + & 1 & = & 1 & \bar{1} & 1, \\ \bar{1} & + & \bar{1} & = & \bar{1} & 1 & \bar{1} \end{array}$$

where $s_k$ is intermediate sum and $c_k$ is carry-overs from the $k$th digit.

We can see that at the mirror-symmetrical summation (subtraction) of the ternary single-digit numerals $a_k + b_k$ of the same sign, there arises the intermediate sum $s_k$ of the opposite sign and the carry-over $c_k$ of the same sign. However, the carry-overs from the $k$th digit spreads simultaneously to the adjacent two digits, namely to the adjacent left, that is, $(k+1)$th digit, and to the adjacent right, that is, $(k-1)$th digit. **Symmetrical appearance of the left and right carry-overs to the side of the adjacent left and right digits is the main peculiarity of the mirror-symmetrical summation (subtraction).**

Table 3.4 describes the functioning of the simplest ternary mirror-symmetrical summator (subtractor), called a *single-digit ternary mirror-symmetrical half-summator (subtractor)*. This half-summator (subtractor) is a combinative logic circuit, which have two ternary inputs $a_k$ and $b_k$ and two ternary outputs $s_k$ and $c_k$ and operates according to Table 3.4 (Fig. 3.2(a)).

As the carry-over from the $k$th digit spreads to the left and to the right digits symmetrically, this means that the full mirror-symmetrical single-digit summator (subtractor) has to have two additional inputs for the carry-overs, entering from the $(k-1)$th and $(k+1)$th digits to the $k$th digit. Thus, the full mirror-symmetrical single-digit summator (subtractor) is a combinative logic circuit, which has 4 ternary inputs and 2 ternary outputs (Fig. 3.2(b)).

Denote by $2\sum$ the mirror-symmetrical single-digit half-summator (subtractor), which has 2 inputs, and by $4\sum$ the mirror-symmetrical single-digit full summator (subtractor), which has 4 inputs.

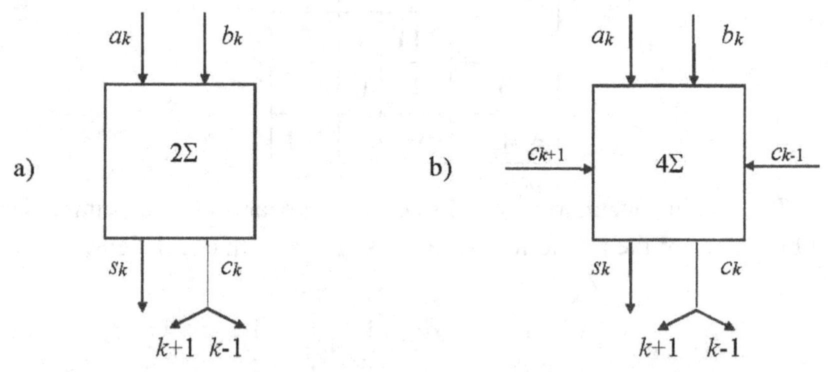


Figure 3.2. Mirror-symmetrical single-digit summators (subtractors): (a) half-summator (subtractor); (b) full summator (subtractor).

Note that we can use the single-digit ternary-symmetrical half-summator in Fig. 2.11 (see Chapter 2) as the single-digit ternary mirror-symmetrical half-summator $2\sum$ in Fig. 3.2(b).

Describe now the logical functioning of the mirror-symmetrical full single-digit summator (subtractor) of the kind 4∑. First of all, we note that the number of all possible 4-digit ternary input combinations of the mirror-symmetrical full summator (subtractor) in Fig. 3.2(b) is equal to $3^4 = 81$. The values of the output variables $s_k$ and $c_k$ are some discrete functions of the algebraic sum $S$ of the input ternary variables $a_k$, $b_k$, $c_{k-1}$, $c_{k+1}$, that is,

$$S = a_k + b_k + c_{k-1} + c_{k+1}. \tag{3.102}$$

The sum (3.102) takes the values from the set {−4, −3, −2, −1, 0, 1, 2, 3, 4}. The functioning of the mirror-symmetrical full summator (subtractor) of the kind 4∑ (Fig. 3.2 (b)) is the following. The summator (subtractor) forms the output ternary code combinations $c_k s_k$ in accordance with the value of the sum (3.102) as follows:

$$-4 = \bar{1}\bar{1}; -3 = \bar{1}0; -2 = \bar{1}1; -1 = 0\bar{1}; 0 = 00; 1 = 01; 2 = 1\bar{1}; 3 = 10; 4 = 11.$$

The lower numerals of such 2-digit ternary representations $c_k s_k$ are the values of the intermediate sums $s_k$ and the higher numerals are the values of the carry-overs $c_k$, which are spreading to the adjacent (left and the right) digits.

Note that the **arithmetical operations of *summation* and *subtraction* are the same arithmetical operations in the ternary mirror-symmetrical arithmetic**. We use the same summators (subtractors) in Fig. 3.2 for their fulfillment.

### 3.4.2. *Examples of mirror-symmetrical summation (subtraction)*

**Example 3.1.** Summarize two ternary mirror-symmetrical numbers 5 + 10:

$$\begin{array}{lccccccccccc}
5 & = & 0 & 1 & \bar{1} & 1. & \bar{1} & 1 & 0 \\
10 & = & 0 & 1 & 1 & 0. & 1 & 1 & 0 \\
S_1 & = & 0 & \bar{1} & 0 & 1. & 0 & \bar{1} & 0 \\
C_1 & & 1 & \leftrightarrow & 1 & & 1 & \leftrightarrow & 1 \\
\hline
15 & = & 1 & \bar{1} & 1 & 1. & 1 & \bar{1} & 1
\end{array}$$

Note that the symbol $\leftrightarrow$ marks the process of carry-over's spreading.

We can see that the summation for this example consists of two steps. The *first step* is to form the first multi-digit intermediate sum $S_1$ and the first multi-digit carry-over $C_1$ according to Table 3.4. The *second step* is summation of the numbers $S_1 + C_1$ according to Table 3.4. As for this case the second multi-digit intermediate carry-over $C_2 = 0$, this means that the summation is over and the sum $S_1 + C_1 = 15$ is the summation result. It is important to emphasize that the summation result

$$15 = 1\bar{1}11.1\bar{1}1 \tag{3.103}$$

is represented in the mirror-symmetrical form.

As we noted above, the important advantage of the ternary mirror-symmetrical numeral system (3.76) is a possibility to summarize all integers (positive and negative) in the "direct" code, that is, without using the notions of "inverse" and "additional" codes.

**Example 3.2.** Subtract the negative ternary mirror-symmetrical number (−24) from the positive ternary mirror-symmetrical number 15. The ternary mirror-symmetrical subtraction is fulfilled as the ternary mirror-symmetrical summation of the ternary mirror-symmetrical numbers (−24) + 15:

$$\begin{array}{rccccccc}
-24= & \bar{1} & \bar{1} & 0 & 1. & 0 & \bar{1} & \bar{1} \\
15= & 1 & \bar{1} & 1 & 1. & 1 & \bar{1} & 1 \\
S_1 & 0 & 1 & 1 & \bar{1}. & 1 & 1 & 0 \\
C_1^1= & & \downarrow & 1 & \leftrightarrow & 1 & \downarrow & \\
C_1^2= & \bar{1} & \leftrightarrow & \bar{1} & & \bar{1} & \leftrightarrow & \bar{1} \\
-9= & \bar{1} & 1 & 1 & \bar{1}. & 1 & 1 & \bar{1}
\end{array}$$

We can see that the ternary mirror-symmetrical summation of numbers (−24) + 15 consists of two steps for the given case. The *first step* is to form the first multi-digit intermediate sum $S_1$ and the first multi-digit carry-over $C_1 = C_1^1 + C_1^2$, according to Table 3.4. The *second step* is to summarize the numbers $S_1 + C_1^1 + C_1^2$. Here we use the functioning rule of the ternary mirror-symmetrical single-digit summator (subtractor) in Fig. 3.2(b). Because for this case the second multi-digit intermediate carry-over $C_2 = 0$, this means that the summation (subtraction) is over and the sum $S_1 + C_1^1 + C_1^2 = -9$ is the result of summation (subtraction). It is important to emphasize that the result of summation (subtraction)

$$-9 = \bar{1}11\bar{1}.11\bar{1} \tag{3.104}$$

is represented in the mirror-symmetrical form.

The most important observation from this example consists of the fact that we used Table 3.4 to summarize the negative number (−24) with a positive number 15. But the summation of two numbers (−24) + 15 is actually reduced to their subtraction. This means that the concept of subtraction as specific arithmetic operation in mirror-symmetrical arithmetic can be abolished. We simple summarize two numbers, which can be of the same sign or opposite signs. For the ternary mirror-symmetrical arithmetic, the signs of the ternary mirror-symmetrical numbers do not have any significance.

### 3.4.3. *Ternary mirror-symmetrical multi-digit summator (subtractor)*

The multi-digit combinative mirror-symmetrical summator (subtractor) (Fig. 3.3), which fulfill the summation (subtraction) of two (2*m* + 1)-digit mirror-symmetrical numbers, is the combinative logic circuit, which consists of the 2*m* + 1 single-digit ternary mirror-symmetrical summators of the kind $4\sum$ (Fig. 3.2(b)).

We can see from Fig. 3.3 that the main peculiarity of the multi-digit ternary mirror-symmetrical summator (subtractor) consists in the fact that the carry-over from each digit is spreading symmetrically to the left and right adjacent digits. Two mirror-symmetrical numbers *A* and *B* enter the multi-digit input of the summator (subtractor). The single-digit summator (subtractor) $4\Sigma_0$ separates the summator (subtractor) into two symmetrical parts: the single-digit summators (subtractors) $4\Sigma_1, 4\Sigma_2, 4\Sigma_3$ for the higher digits and the single-digit summators (subtractors) $4\Sigma_{-1}, 4\Sigma_{-2}, 4\Sigma_{-3}$ for the lower digits.

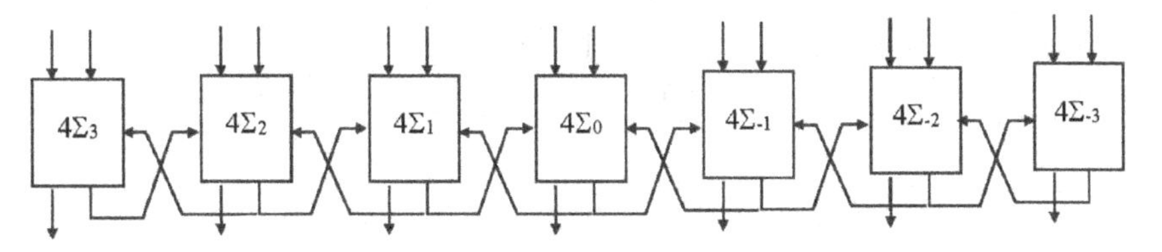


Figure 3.3. The ternary mirror-symmetrical multi-digit summator (subtractor).

Note that the ternary multi-digit mirror-symmetrical summator (subtractor) in Fig. 3.3 fulfils the operations of ternary mirror-symmetrical summation and subtraction of the ternary multi-digit mirror-symmetrical numbers. This means that the ternary multi-digit mirror-symmetrical summation of the two positive mirror-symmetrical numbers with equal signs 5 + 10 (Example 3.1) and two mirror-symmetrical numbers with different signs (−24) + 15 (Example 3.2) can be fulfilled on the same ternary mirror-symmetrical multi-digit summator, which is also ternary mirror-symmetrical multi-digit subtractor.

### 3.4.4. *"Swing"-phenomenon*

Summarize now two equal numbers 5 + 5, represented in the ternary mirror-symmetrical numeral system (3.76):

$$
\begin{array}{ccccccccc}
5 & = & 0 & 1 & \bar{1} & 1. & \bar{1} & 1 & 0 \\
5 & = & 0 & 1 & \bar{1} & 1. & \bar{1} & 1 & 0 \\
\hline
 & & 0 & \bar{1} & 1 & \bar{1}. & 1 & \bar{1} & 0 \\
 & & & \downarrow & 1 & \leftrightarrow & 1 & \downarrow & \\
 & & 1 & \leftrightarrow & 1 & & 1 & \leftrightarrow & 1 \\
 & & & \bar{1} & \leftrightarrow & \bar{1} & \downarrow & & \\
 & & & & & \bar{1} & \leftrightarrow & \bar{1} & \\
\hline
 & & 1 & 1 & 0 & 0. & 0 & 1 & 1 \\
 & & & & \bar{1} & \leftrightarrow & \bar{1} & & \\
 & & & 1 & \leftrightarrow & 1 & \downarrow & & \\
 & & & \downarrow & & 1 & \leftrightarrow & 1 & \\
 & & \bar{1} & \leftrightarrow & \bar{1} & & \bar{1} & \leftrightarrow & \bar{1} \\
\hline
 & & 0 & \bar{1} & 1 & \bar{1}. & 1 & \bar{1} & 0 \\
 & & & & 1 & \leftrightarrow & 1 & & \\
 & & & \bar{1} & \leftrightarrow & \bar{1} & \downarrow & & \\
 & & & \downarrow & & \bar{1} & \leftrightarrow & \bar{1} & \\
 & & 1 & \leftrightarrow & 1 & & 1 & \leftrightarrow & 1
\end{array}
$$

As it follows from this example, we have found a special kind of summation (subtraction) called a *"swing"-phenomenon* [28]. If we continue the summation process in the above example, then, by starting from some step, the process of the carry-overs formation turns out to be repetitive and hence the summation becomes infinite. The *"swing"-phenomenon* is a variety of the "races," which arise in digit automatons, when elements are switching over.

To eliminate the *"swing"-phenomenon*, we can use the following effective "technical" method [28]. The *"swing"-phenomenon* arises in the ternary mirror-symmetrical summator (subtractor) in Fig. 3.3 because the carry-overs come simultaneously on some summation step from the two adjacent single-digit summators (subtractors) with the odd indices ($k = \pm1, \pm3, \pm5, \ldots$). To eliminate the *"swing"-phenomenon*, on the first summation step only the summators (subtractors) with the *even* indices ($k = 0, \pm2, \pm4, \ldots$) form the intermediate sums and corresponding carry-overs to the single-digit summators (subtractors) with the *odd* indices ($k = \pm1, \pm3, \pm5, \ldots$). Then at the second step of the ternary mirror-symmetrical summation (subtraction), the carry-overs, which were formed at the first step, are summarized with the corresponding ternary numerals of the odd digits of the summarized numbers. Thanks to such approach the *"swing"-phenomenon* is eliminated.

Let us demonstrate now the above method of the elimination of the *"swing"-phenomenon* at the summation of the numbers 5 + 5:

| | | | | | | | | |
|---|---|---|---|---|---|---|---|---|
| 5 | = | 0 | 1 | $\bar{1}$ | 1. | $\bar{1}$ | 1 | 0 |
| 5 | = | 0 | 1 | $\bar{1}$ | 1. | $\bar{1}$ | 1 | 0 |
| $S_1$ | = | | $\bar{1}$ | | $\bar{1}$. | | $\bar{1}$ | |
| | | | ↓ | | ↓ | | ↓ | |
| $C_1^1$ | | | ↓ | 1 | ↔ | 1 | ↓ | |
| $C_1^2$ | | 1 | ↔ | 1 | | 1 | ↔ | 1 |
| 10 | = | 1 | $\bar{1}$ | 0 | $\bar{1}$. | 0 | $\bar{1}$ | 1 |

The first step of the mirror-symmetrical summation is to summarize all the input ternary numerals with *even* indices (2, 0, −2). The ternary numerals of all digits with *odd* indices (3, 1, −1, −3) are delayed at the first step. The second step is summation of all the carry-overs, which arise at the first step, with the input ternary numerals of the digits with *odd* indices. It is important to emphasize that the summation result

$$10 = 1\bar{1}0\bar{1}.0\bar{1}1 \qquad (3.105)$$

is represented in the mirror-symmetrical form.

The analysis of all the above examples of the ternary mirror-symmetrical summation (subtraction) show that both the final summation (subtraction) results and all the intermediate summation (subtraction) results are ternary mirror-symmetrical numbers, that is, the property of *the mirror symmetry is an invariant about the mirror-symmetrical summation (subtraction)*. This means that mirror-symmetrical *summation (subtraction)* possesses the important mathematical property of "*mirror symmetry,*" which can be used for checking the ternary mirror-symmetrical summator (subtractor) in Fig. 3.3.

## 3.5. The ternary mirror-symmetrical multiplication and division

### 3.5.1. *Mirror-symmetrical multiplication*

The following trivial identity for the golden ratio powers underlies the mirror-symmetrical multiplication:

$$\Phi^{2n} \times \Phi^{2m} = \Phi^{2(n+m)}. \qquad (3.106)$$

The rule of the mirror-symmetrical multiplication of two single-digit ternary mirror-symmetrical numbers is given in Table 3.5.

Table 3.5. Ternary mirror-symmetrical multiplication.

| $a_k \times b_k$ | $\bar{1}$ | 0 | 1 |
|---|---|---|---|
| $\bar{1}$ | 1 | 0 | $\bar{1}$ |
| 0 | 0 | 0 | 0 |
| 1 | $\bar{1}$ | 0 | 1 |

Comparison of Table 2.2 (Chapter 2) and Table 3.5 shows that the rule of the classic ternary-symmetrical multiplication (see Chapter 2) coincides with the ternary mirror-symmetrical multiplication. The ternary mirror-symmetrical multiplication is fulfilled in the "direct" code. For technical realization of the ternary mirror-symmetrical multiplication the same ternary single-digit multiplier as in Fig. 2.11 can be used (Chapter 2).

The general algorithm of the multiplication of two ternary multi-digit mirror-symmetrical numbers is reduced to the formation of the partial products in accordance with Table 3.5 and their summation (subtraction) in accordance with the rule of the ternary mirror-symmetrical summation-subtraction (Table 3.4).

**Example 3.3.** Multiply the negative ternary mirror-symmetrical number $-6=\bar{1}01.0\bar{1}$ by the positive ternary mirror-symmetrical number $2=1\bar{1}.1$:

$$\begin{array}{ccccccc}
 & \bar{1} & 0 & 1. & 0 & \bar{1} & \\
 & & 1 & \bar{1}. & 1 & & \\
\hline
 & & \bar{1} & 0. & 1 & 0 & \bar{1} \\
 & 1 & 0 & \bar{1}. & 0 & 1 & \\
\bar{1} & 0 & 1 & 0. & \bar{1} & & \\
\hline
\bar{\bar{1}} & 1 & 0 & \bar{\bar{1}}. & 0 & 1 & \bar{\bar{1}}
\end{array}$$

The multiplication result in Example 3.3 is formed as the sum of three partial products. The first partial product $\bar{1}0.10\bar{1}$ is the result of multiplication of the ternary mirror-symmetrical multiplier $-6=\bar{1}01.0\bar{1}$ by the lowest positive numeral 1 of the ternary mirror-symmetrical multiplier $2=1\bar{1}.1$, the second partial product $10\bar{1}.01\ 1$ is the result of the multiplication of the same number $-6=\bar{1}01.0\bar{1}$ by the middle negative numeral $\bar{1}$ of the number $2=1\bar{1}.1$, and, finally, the third partial product $\bar{1}010.\bar{1}$ is the result of the multiplication of the same number $-6=\bar{1}01.0\bar{1}$ by the higher positive numeral 1 of the number $2=1\bar{1}.1$.

Note that the product $-12=\bar{1}10\bar{1}.01\bar{1}$ is represented in the mirror-symmetrical form! As its higher digit is a negative numeral $\bar{1}$, it follows from this that the product $(-12)$ is a negative ternary mirror-symmetrical number.

### 3.5.2. *Mirror-symmetrical division*

The ternary mirror-symmetrical division is fulfilled in accordance with the division rule of the classic ternary-symmetrical numeral system (see Chapter 2). The general algorithm of the ternary mirror-symmetrical division is reduced to the sequential subtraction of the shifted divisor, which is multiplied by the next ternary numeral of the quotient.

### 3.5.3. *The main arithmetical advantages of the ternary mirror-symmetrical arithmetic*

We can point on the number of the important advantages of the ternary mirror-symmetrical arithmetic from the "technical" point of view:

(1) The mirror-symmetrical subtraction is the same arithmetic operation as the mirror-symmetrical summation. The mirror-symmetrical summation (subtraction) is fulfilled by means of one and the same mirror-symmetrical summator (subtractor) in Fig. 3.3 in the "direct" code, that is, without using the notions of the *inverse* and *additional* codes.

(2) The sign of the summarized or subtracted numbers is defined automatically, because it coincides with the sign of the higher significant ternary numeral of the ternary mirror-symmetric representation of the summation (subtraction) result.

(3) The summation (subtraction) results are always represented in the mirror-symmetrical form that allows checking a process of the ternary mirror-symmetrical summation (subtraction) according to the property of "*mirror-symmetry.*"

(4) The mirror-symmetrical multiplication is reduced to the mirror-symmetrical summation (subtraction). The ternary mirror-symmetrical multiplication can be fulfilled over ternary mirror-symmetrical numbers of equal signs or different signs in the "direct" code, that is, without the use of the notions of the *inverse* and *additional* codes.

(5) The sign of the results of the mirror-symmetrical multiplication is defined automatically because it coincides with the sign of the higher significant ternary numeral of the ternary mirror-symmetrical representation of the result of the mirror-symmetrical multiplication.

(6) The results of the mirror-symmetrical multiplication are always represented in the mirror-symmetrical form that allows checking a process of the ternary mirror-symmetrical multiplication.
(7) The operation of the ternary mirror-symmetrical division is more complicated arithmeticaly than that of the ternary mirror-symmetrical summation, subtraction and multiplication. By its complexity, the operation of the ternary mirror-symmetrical division is comparable to the same operation in the ternary-symmetrical numeral system, used in the ternary computer "Setun."

## 3.6. Technical realizations of the ternary mirror-symmetrical arithmetical devices

### 3.6.1. *General information*

By comparing the classical ternary-symmetrical arithmetic, based on the following property of the ternary numbers:

$$1+1=3^n+3^n=3^{n+1}-3^n=1\bar{1}, \qquad (3.107)$$

with the ternary mirror-symmetrical arithmetic, based on the following property of the golden ratio

$$1+1=\Phi^{2n}+\Phi^{2n}=\Phi^{2(n+1)}-\Phi^{2n}+\Phi^{2(n-1)}=1\bar{1}1, \qquad (3.108)$$

we can see that there is similarity between (3.107) and (3.108) from the point of view of technical realization. As it is known, the rule of the intermediate sum and carry-over formation at the summation (subtraction) of two single-digit ternary-symmetrical numbers follows from (3.107). This rule consists of the formation of the intermediate sum $\bar{1}$ and the carry-over 1, which is spread to the adjacent higher digit.

The identity (3.108) gives the rule of the formation of the intermediate sum and the carry-over at the summation (subtraction) of the two single-digit ternary numerals $\{\bar{1},0,1\}$. In accordance with (3.108), there are formed the intermediate sum $\bar{1}$ and the carry-over 1, which is spread to the left and right adjacent digits relatively to the initial digit. We can see that *the rules of the formation of intermediate sum and carry-over at the summation-subtraction of the single-digit ternary-symmetrical and ternary mirror-symmetrical numerals coincide*. The distinction consists only in spreading the carry-overs. For the case (3.107) the carry-over is spreading to the left, that is, to the side of the higher digit, for the case (3.108) the carry-over is spreading symmetrically relative to the initial digit, that is, to the left and right digits simultaneously.

It follows from this consideration a very important technical conclusion: *the logic circuits for the realization of the single-digit transformation of the ternary-symmetrical arithmetic and ternary mirror-symmetrical arithmetic are identical*. In particular, we can use without any change at the technical realization of the ternary mirror-symmetrical arithmetic the same logic circuits of the ternary-symmetrical arithmetic (see Chapter 2).

### 3.6.2. *The ternary mirror-symmetrical accumulator*

In the above we have developed the multi-digit ternary mirror-symmetrical summator (subtractor) (Fig. 3.3). This summator (subtractor) is the basis for the ternary mirror-symmetrical accumulator, which is a central device of the ternary mirror-symmetrical processor (Fig. 3.4).

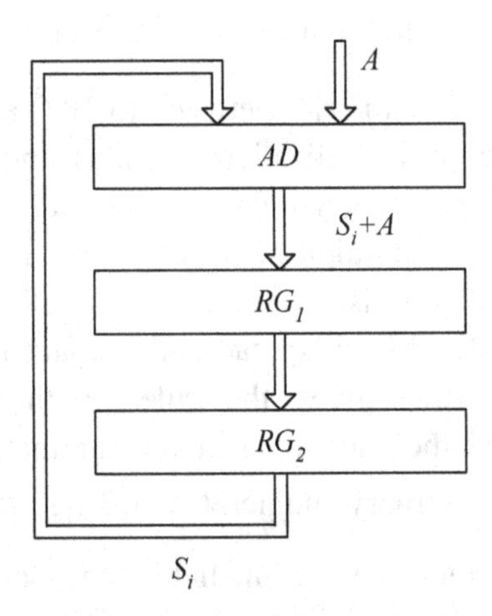


Figure 3.4. Ternary mirror-symmetrical accumulator.

The accumulator in Fig. 3.4 has a traditional structure and consists of the multi-digit ternary mirror-symmetrical summator (subtractor) *AD*, the intermediate ternary register $RG_1$, which memorizes the sum $S_1 + A$, arising at the *AD*-output, and the accumulating ternary register $RG_2$.

The ternary mirror-symmetrical accumulator in Fig. 3.4 is a universal device of the ternary mirror-symmetrical processor and underlies the other ternary mirror-symmetrical devices. By using the additional devices, we can construct the following ternary mirror-symmetrical devices:

(a) If we add to the input *A* the positive (1) or negative $(\bar{1})$ ternary numerals sequentially, we turn the accumulator into the summing or subtracting ternary mirror-symmetrical counter.
(b) If we add the device for the formation of the partial products $A \times b_i \Phi^{2i}$ before the input *A*, we turn the accumulator into the multiplier.
(c) As the mirror-symmetrical division is reduced to the shift of the divisor and its subtraction from the dividend, the mirror-symmetrical accumulator in Fig. 3.4 can be used for the design of the ternary mirror-symmetrical divider.

## 3.7. Matrix and pipeline mirror-symmetrical arithmetical unit

### 3.7.1. *Matrix mirror-symmetrical summator (subtractor)*

It is well known that the digital signal processors put forward high demands to the speed of arithmetical devices. The different special structures (matrix, pipeline, etc.) are elaborated for this purpose. We can show in this section that the ternary mirror-symmetrical arithmetic contains the interesting possibilities for designing the fast ternary arithmetical processors.

Let us consider now the matrix multi-digit ternary mirror-symmetrical summator (subtractor) (Fig. 3.5). Each cell of the ternary mirror-symmetrical matrix summator (subtractor) in Fig. 3.5 is a single-digit ternary-symmetrical full summator (subtractor), which have 4 inputs and 2 outputs (see Fig. 3.2(b)).

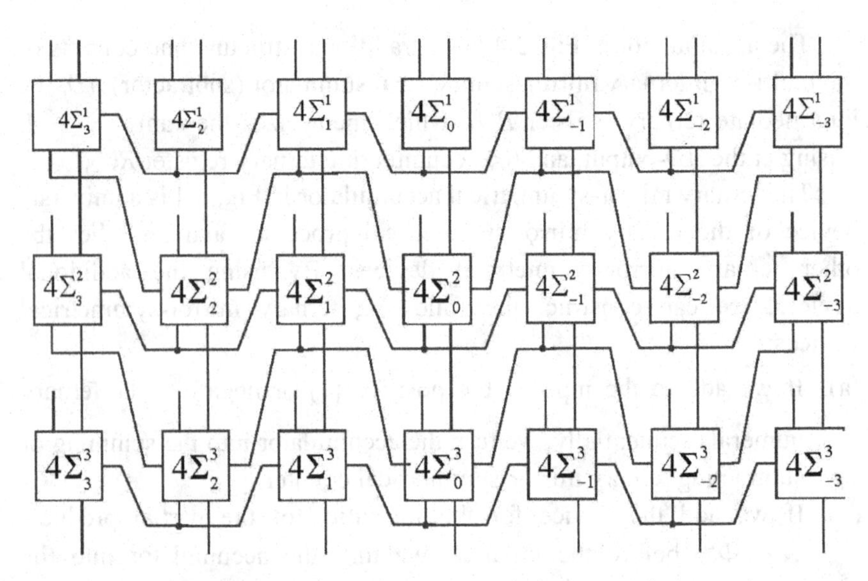


Figure 3.5. Matrix ternary mirror-symmetrical summator (subtractor).

The matrix ternary mirror-symmetrical summator (subtractor) in Fig. 3.5 consists of 21 single-digit full summators (subtractors), which are arranged in the form of the 7 × 3-matrix. Each ternary single-digit summator (subtractor) has a designation $4\Sigma_i^k$, where the number 4 means that the summator (subtractor) $4\Sigma_i^k$ has 4 ternary inputs, the indices *i* and *k* in the summator (subtractor) $4\Sigma_i^k$ mean that the summator (subtractor) $4\Sigma_i^k$ refers to the *i*th digit of the ternary mirror-symmetrical code (3.76) and the summator (subtractor) are placed in the *k*th row of the matrix summator (subtractor) in Fig. 3.5.

The inputs of the single-digit summator (subtractor)

$$4\Sigma^1_3, 4\Sigma^1_2, 4\Sigma^1_1, \Sigma^1_0, 4\Sigma^1_{-1}, 4\Sigma^1_{-2}, 4\Sigma^1_{-3}$$

of the first row form the multi-digit input of the matrix ternary-symmetric summator (subtractor) in Fig. 3.5. The output of the intermediate sum of each single-digit summator (subtractor) is connected to the corresponding input of the next single-digit summator (subtractor) of the same column.

The outputs of the intermediate sum of the single-digit summator (subtractor)

$$4\Sigma_3^3, 4\Sigma_2^3, 4\Sigma_1^3, \Sigma_0^3, 4\Sigma_{-1}^3, 4\Sigma_{-2}^3, 4\Sigma_{-3}^3$$

of the last row form the multi-digit output of the matrix mirror-symmetrical summator (subtractor).

The main peculiarity of the matrix mirror-symmetrical summator (subtractor) in Fig. 3.5 consists in a special organization of the connections between the carry-over outputs of the single-digit summators (subtractors) and the inputs of the adjacent single-digit summators (subtractors). The carry-over outputs of all single-digit summators (subtractors) with the *even* lower indices (2, 0, −2) are connected to the corresponding inputs of the adjacent single-digit summators (subtractors), which are placed in the same row, but the carry-over outputs of all the single-digit summators (subtractors) with the *odd* lower indices (3, 1, −1, −3) are connected with the corresponding inputs of the adjacent single-digit summators (subtractors), which are placed in the lower row. Note that such organization of the carry-over connections allows eliminating the above "swing" phenomenon.

Consider the operation of the matrix mirror-symmetric summator (subtractor) on the example of summing two equal ternary mirror-symmetrical numbers:

$$A = 0111.110 \text{ and } B = 0111.110.$$

The summation is fulfilled in two stages. Each stage is fulfilled by means of one row of the single-digit summators (subtractors) and consists of two steps.

## ***The first stage***

In accordance with Fig. 3.5 the *first step* of the first stage consists in the following. The single-digit summators (subtractors) of the first row with the *even* lower indices ($4\Sigma_2^1$, $4\Sigma_0^1$, $4\Sigma_{-2}^1$) form the intermediate sums, which enter the inputs of the second row of summators (subtractors), and the carry-overs, which enter the corresponding inputs of the single-digit summators (subtractors) with the *odd* lower indices of the first row ($4\Sigma_3^1$, $4\Sigma_1^1$, $4\Sigma_{-1}^1$, $4\Sigma_{-3}^1$). Such transformation of the code information can be represented in the following form:

$$\begin{array}{ccccccc}
0&1&1&1.&1&1&0\\
0&1&1&1.&1&1&0\\ \hline
&\bar{1}&&\bar{1}&&\bar{1}&\\
&\downarrow&1&\leftrightarrow&1&\downarrow&\\
1&\leftrightarrow&1&&1&\leftrightarrow&1
\end{array}$$

Hence, the first step is the formation of the intermediate sums and the carry-overs on the outputs of the single-digit summators (subtractors) with the *even* lower indices (2, 0, −2).

At the *second step* of the first stage the single-digit summators (subtractors) with the *odd* lower indices (3, 1, −1, −3) go into action. In accordance with the entered carry-overs they form the intermediate sums and the carry-overs, entering the single-digit summators (subtractors) of the lower row, that is,

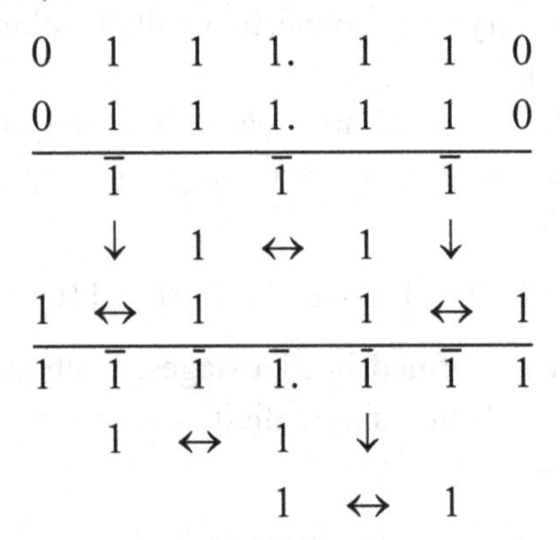

The first stage is over. We can see that the results of the first stage are some intermediate sum and some carry-overs, entering the summators (subtractors) of the lower row.

***<u>The second stage</u>***

The single-digit summators (subtractors) of the second row with the *even* lower indices ($4\Sigma_2^2$, $4\Sigma_0^2$, $4\Sigma_{-2}^2$) form the intermediate sums, entering the corresponding inputs of the lower row summators (subtractors) and the carry-overs, entering the corresponding inputs of the same row summators (subtractors) with the *odd* lower indices ($4\Sigma_3^2$, $4\Sigma_1^2$, $4\Sigma_{-1}^2$, $4\Sigma_{-3}^2$), that is,

$$\begin{array}{ccccccc} 1 & \bar{1} & \bar{1} & \bar{1}. & \bar{1} & \bar{1} & 1 \\ & 1 & \leftrightarrow & 1 & \downarrow & & \\ & & & 1 & \leftrightarrow & 1 & \\ \hline 1 & 0 & \bar{1} & 1. & \bar{1} & 0 & 1 \end{array}$$

As all the carry-overs, which are formed at this stage, became equal to 0, this means that the summation (subtraction) process is over at the second stage (this is true only for the considered case). The obtained sum enters the inputs of the lower row summators (subtractors) $4\Sigma_3^3$ - $4\Sigma_{-3}^3$ and then appears on the output of the matrix summators (subtractors).

### 3.7.2. *The pipeline mirror-symmetrical summator (subtractor)*

There are two ways to extend the functional possibilities of matrix mirror-symmetrical summator (subtractor) in Fig. 3.5. If we place the ternary registers, which consist of the flip-flop-flap's (see Fig. 2.14(b)) between the adjacent rows of the single-digit summators (subtractors), then the above matrix summator (subtractor) turns into the *pipeline* ternary mirror-symmetrical summator (subtractor). In fact, the code information from the preceding rows of the single-digit summators (subtractors) is memorized in the corresponding ternary registers and the preceding row of the summators (subtractors) is ready for further processing. Then, the summators (subtractors) of the lower row process the code information, enter the lower row of the single-digit summators (subtractors), and simultaneously the top row of the single-digit summators (subtractors) starts processing new input code information. This means that since the given moment we get the sums of the numbers $A_1 + B_1, A_2 + B_2, \ldots, A_n + B_n$, entering the summator (subtractor) during the time period $2\Delta\tau$, where $\Delta\tau$ is the delay time of the single-digit summator (subtractor).

### 3.7.3. *The pipeline ternary mirror-symmetrical multiplier*

The other way to extend functional possibilities of the pipeline summator (subtractor) consists of the following. We can see in Fig. 3.5 that each single-digit summator (subtractor) of the lower rows has a "free" input. We can use these inputs as new multi-digit inputs of the pipeline summator (subtractor). By using these multi-digit inputs, we can turn the pipeline summator (subtractor) into the *pipeline multiplier*. In this case the mirror-symmetric multiplication of two ternary mirror-symmetrical numbers $A(1) \times B(1)$ is fulfilled in the following manner. The first row of the single-digit summator (subtractor) summarizes the first two partial products $\mathrm{P}_1^1 + \mathrm{P}_2^1$. This code information enters the second row of the single-digit summators (subtractors). If we send the 3rd partial product $\mathrm{P}_3^1$ to the "free" input of the second row, we get the sum $\mathrm{P}_1^1 + \mathrm{P}_2^1 + \mathrm{P}_3^1$ on the outputs of the second row. In this case, the first row starts to sum the first two partial products of the next pair of multiplied numbers $A(2) \times B(2)$. The "free" input of the 3rd row is used to accept the next partial product $\mathrm{P}_4^1$ of the first pair of the multiplied numbers $A(1) \times B(1)$, etc. We can see that the pipeline summator (subtractor) in Fig. 3.5 allows multiplying many ternary mirror-symmetrical numbers in the pipeline regime. In this connection the multiplication speed is determined by the time $2\Delta\tau$, where $\Delta\tau$ is the delay time of the single-digit summator (subtractor).

## 3.8. Conclusions to Chapter 3

### 3.8.1. *The ternary mirror-symmetrical arithmetic is a synthesis of Bergman's system and Brusentsov's ternary principle*

The *Bergman's system* [10] is the first achievement, which underlies the ternary mirror-symmetrical arithmetic. The *Brusentsov's ternary principle*, embodied in the ternary computer "Setun" (Moscow University, 1957, Principal Designer Nikolay Brusentsov), is the second achievement in modern computer science, which was used in the ternary mirror-symmetrical arithmetic.

The ternary mirror-symmetrical numeral system with the irrational base $\Phi^2 = \frac{3+\sqrt{5}}{2}$ (the square of the golden ratio) [28] retains all the major advantages of *Bergman's system* and the classic *ternary-symmetrical numeral system*. Its main advantage in comparison with the conventional ternary-symmetrical numeral system is an original way to check all the major informational transformations according to the "*principle of mirror symmetry.*" Therefore, the author believes that the problem of the implementation of the ternary mirror-symmetrical arithmetic into modern computer technology may be a matter of not so distant future.

### 3.8.2. *Supporting of Prof. Donald Knuth*

The difficulties, associated with the recognition of new ideas and concepts in science and mathematics, are well known. The fate of the non-Euclidean geometry, developed by young Russian mathematician Nikolay Lobachevsky (Kazan University), is the classic example. This outstanding mathematical discovery was evaluated very negatively by the Russian mathematical community (negative review by academician M.V. Ostrogradsky). But thanks to the support of the outstanding German mathematician Friedrich Gauss, Lobachevsky's geometry had been recognized by the international mathematical community, and later the Russian mathematical community was forced to recognize Lobachevsky's geometry as the outstanding mathematical achievement of the 19th century [1].

In this regard, it is appropriate to evaluate the role of the prominent American mathematician and world expert in computer science Prof. Donald Knuth in the development of numeral systems with irrational bases.

Figure 3.6. American mathematician and world expert in computer science Prof. Donald Knuth.

Unfortunately, mathematicians and computer experts of that time were not able to appreciate the *Bergman's system* as the outstanding mathematical discovery. Prof. Donald Knuth was the only exception. He posted a link to Bergman's article [10] in his best-selling book *The Art of Computer Programming* [72] and in this manner he had supported Bergman's mathematical discovery. Thanks to Donald Knuth's book [72], the author of the present book had learned Bergman's article [10] which influenced the author's research in the theory of numeral systems with irrational bases and their applications.

In 2002 the well-known international magazine *The Computer Journal* (British Computer Society) had published author's article "*Brusentsov's Ternary Principle, Bergman's Number System and Ternary Mirror-symmetrical Arithmetic*" [28]. Prof. Donald Knuth was the first prominent scientist, who has responded to this publication. In his letter, Prof. Knuth appreciated highly the article [28], and told about his intention to give a description of the ternary mirror-symmetrical arithmetic in the new edition of the book [72]. This letter of the famous scientist is the highest award for author's researches in the field of new computer arithmetic's.

### 3.8.3. *Supporting Nikolay Brusentsov*

Brusentsov's works on the creation of the first ternary computer "Setun" [71] became the author's motivation for the creation of new ternary mirror-symmetrical arithmetic, based on the golden ratio.

On May 29, 2003 the author made a speech at the session of the prestigious scientific seminar "*Geometry and Physics*" (Seminar leader Prof. Yuri Mikhailov, Department of Theoretical Physics, Moscow University). During the speech the author had met with the patriarch of the Soviet computer science Nikolay Brusentsov.

Figure 3.7. Nikolay Brusentsov and Alexey Stakhov (Moscow University, May 29, 2003).

Nikolay Brusentsov participated in the discussion of author's speech and said the following:

*"Fibonacci series and its generalizations, the golden ratio and following from it the binary Φ-code with irrational base* $\Phi = \frac{1+\sqrt{5}}{2}$ *and the ternary mirror-symmetrical code with irrational base* $\Phi = \frac{3+\sqrt{5}}{2}$ *are fundamental components of the structure of the world order. The*

*impressive Stakhov's results will contribute to overcome the crisis in the foundations of modern science."*

The author thanks Nikolay Brusentsov for the appreciation of his work in the field of numeral systems with irrational bases.

### 3.8.4. *Evaluation of Bergman's system and its applications*

As is known, new scientific ideas do not always arise there, where they are expected. Apparently, *Bergman's system* [10] is one of the most unprecedented scientific discoveries in the history of science and mathematics. First of all, it is impressive that this discovery in its origins goes back to the Babylonian positional numeral system with base 60 that arose at the stage of mathematics origin. That is, Bergman's mathematical discovery [10] returns mathematics to the initial period of its development, when the numeral systems and rules of arithmetic operations were one of the most important parts of mathematics (Babylon and ancient Egypt) [2].

However, the greatest impression is the fact that a new scientific discovery in the theory of numeral systems was made by 12-year-old American wunderkind George Bergman. This is really an unprecedented event in the history of science and mathematics. The mathematical formula (3.1) for the *Bergman's system* looks so simple that it is difficult to believe that *Bergman's system* is one of the largest modern mathematical discoveries, which is of fundamental interest for history of mathematics, number theory and computer science. In this regard, one can compare *Bergman's system* with the discovery of *incommensurable segments*, made in Pythagoras' scientific school. The proof of the *incommensurability* of the diagonal and the side of the square is so simple that any amateur of mathematics can prove this without any difficulties. However, this mathematical discovery still causes delight, since it was this discovery that was the turning event in the development of mathematics and led to the introduction of irrational numbers, without which it is difficult to imagine the existence of mathematics. Time will show how fair the above comparison of *Bergman's system* with the discovery of *incommensurable segments*.

*Bergman's system* contains many interesting ideas that can lead to new results in mathematics and applied sciences, in particular, computer science. Among these results we can name the "golden" number theory [33], from which new properties of natural numbers follow, and the "golden" ternary mirror-symmetrical arithmetic [28], which is of great interest for the future of computer science.

## Chapter 4

# Fibonacci *p*-Codes and the Concept of Fibonacci Computers

## 4.1. Fibonacci *p*-codes

### 4.1.1. *A definition of the Fibonacci p-codes*

In 1972, the author of this book defended his doctoral thesis on *"Synthesis of Optimal Algorithms for Analog-Digital Conversion"* [11]. On the basis of this dissertation the author wrote the book "*Introduction into Algorithmic Measurement Theory*" [14], devoted to the substantiation of the theory of the so-called *Fibonacci p-codes*:

$$N = a_n F_p(n) + a_{n-1} F_p(n-1) + \ldots + a_i F_p(i) + \ldots + a_1 F_p(1), \quad (4.1)$$

where $N$ is positive integer, $a_i \in \{0, 1\}$ is a binary numeral of the $i$th digit of code (4.1); $n$ is the digit number of code (4.1).

Here

$$\{F_p(1), F_p(2), \ldots, F_p(i), \ldots, F_p(n)\} \quad (4.2)$$

are the weights of code (4.1), $F_p(i)(i = 1, 2, 3, \ldots, n)$ are Fibonacci $p$-numbers, calculated in accordance with the recurrent relation (1.57) $F_p(n) = F_p(n-1) + F_p(n-p-1)$ at the seeds (1.58) $F_p(1) = F_p(2) = \ldots = F_p(p+1) = 1$.

### 4.1.2. *Partial cases of the Fibonacci p-codes*

Note that the Fibonacci $p$-codes (4.1) include infinite number of the different positional "binary" representations of positive integers because every $p$ originates its own Fibonacci $p$-code ($p = 0, 1, 2, 3, \ldots$).

In particular, for the case $p = 0$ Fibonacci $p$-code (4.1) is reduced to the classic binary code:

$$N = a_n 2^{n-1} + a_{n-1} 2^{n-2} + \ldots + a_i 2^{i-1} + \ldots + a_1 2^0. \qquad (4.3)$$

For the case $p = 1$ Fibonacci $p$-code (4.1) is reduced to the classic Fibonacci code, named *Fibonacci 1-code*:

$$N = a_n F_n + a_{n-1} F_{n-1} + \ldots + a_i F_i + \ldots + a_1 F_1, \qquad (4.4)$$

where $F_i = F_{i-1} + F_{i-2}$; $F_1 = F_2 = 1$ ($i = 1, 2, 3, \ldots, n$) are the classic Fibonacci numbers.

The abridged representations of Fibonacci $p$-code (4.1), the classic binary code (4.3) and the Fibonacci 1-code (4.4) have one and the same form:

$$N = a_n a_{n-1} \ldots a_i \ldots a_1, \qquad (4.5)$$

which is named *Fibonacci representations*.

Consider now the partial case $p = \infty$. For this case every Fibonacci $p$-number is equal to 1, that is, for any integer $i = 1, 2, 3, \ldots, n$ we have: $F_p(i) = 1$. Then, for this case the sum (4.1) takes the form of the so-called *unitary code*:

$$N = \underbrace{1 + 1 + \ldots + 1}_{N}. \qquad (4.6)$$

Hence, Fibonacci $p$-codes, given by (4.1), are a very wide generalization of binary system (4.3) and Fibonacci 1-code (4.4), which are the partial cases of Fibonacci $p$-codes (4.1) for the cases $p = 0$ and $p = 1$, respectively. On the other hand, Fibonacci $p$-code (4.1) for the case $p = \infty$ is reduced to the *unitary code* (4.6).

It follows from the above reasoning that a number of Fibonacci $p$-codes, given by (4.1), is theoretically infinite. However, they in general have different practical significance. For Fibonacci $p$-codes (4.1) the *principle of simplicity of technical realization* plays the same important role as for the case of the canonical systems (2.2) and the symmetrical systems (2.6). The binary system (4.3) has the most practical significance among the canonical systems (2.2) and the ternary-symmetrical system (2.8) has the most practical significance among the symmetrical systems (2.6).

The situation is similar for the Fibonacci $p$-codes (4.1). The Fibonacci $p$-codes, which corresponds to the cases $p = 0$ (the binary code (4.3)) and $p = 1$ (the classic Fibonacci code (4.4)), are of great practical significance. As it is showed below, the classic binary code (4.3) has "zero" redundancy and doesn't have any practical interest for error detection, our main task in Chapter 4 consists in studying the properties and applications of the simplest redundant Fibonacci 1-code (4.4) and its benefits in comparison to the classic binary code (4.3). The Fibonacci 1-code (4.4), based on the classic Fibonacci numbers, is the main focus of our research in Chapter 4.

### 4.1.3. *A range of number representation in the Fibonacci p-code*

Consider the set of $n$-digit binary words. A number of them are equal to $2^n$. For the classic binary code (4.3) ($p = 0$) the mapping of $n$-digit binary words onto the set of the natural numbers has the following peculiarities:

(a) *Uniqueness of mapping*. This means that for the case $n \rightarrow \infty$ there is one-to-one correspondence between natural numbers and sums (4.3), that is, each positive integer $N$ has the only binary representation in the form (4.5).

(b) For the given $n$ by using the binary code (4.3) we can represent all positive integers in the range from 0 to $2^n - 1$, that is, the range of number representation is equal to $2^n$.

(c) The minimal number 0 and the maximal number $2^n - 1$ have the following binary representations in the binary code (4.3):

| | | |
|---|---|---|
| 0 | = | 00...0 |
| $2^n - 1$ | = | 11...1 |

For the Fibonacci $p$-codes (4.1) the mapping of the $n$-digit Fibonacci representations (4.5) onto positive integers has distinct peculiarities for the case $p > 0$.

Let $n = 5$. Then for the case $p = 1$ the mapping of the 5-digit Fibonacci 1-code (4.4) onto the natural numbers has the form, represented in Table 4.1.

Table 4.1. Mapping the Fibonacci 1-code representations onto natural numbers.

| $FR/W$ | 5 | 3 | 2 | 1 | 1 | $N$ | $FR/W$ | 5 | 3 | 2 | 1 | 1 | $N$ |
|---|---|---|---|---|---|---|---|---|---|---|---|---|---|
| $A_0$ | 0 | 0 | 0 | 0 | 0 | 0 | $A_{16}$ | 1 | 0 | 0 | 0 | 0 | 5 |
| $A_1$ | 0 | 0 | 0 | 0 | 1 | 1 | $A_{17}$ | 1 | 0 | 0 | 0 | 1 | 6 |
| $A_2$ | 0 | 0 | 0 | 1 | 0 | 1 | $A_{18}$ | 1 | 0 | 0 | 1 | 0 | 6 |
| $A_3$ | 0 | 0 | 0 | 1 | 1 | 2 | $A_{19}$ | 1 | 0 | 0 | 1 | 1 | 7 |
| $A_4$ | 0 | 0 | 1 | 0 | 0 | 2 | $A_{20}$ | 1 | 0 | 1 | 0 | 0 | 7 |
| $A_5$ | 0 | 0 | 1 | 0 | 1 | 3 | $A_{21}$ | 1 | 0 | 1 | 0 | 1 | 8 |
| $A_6$ | 0 | 0 | 1 | 1 | 0 | 3 | $A_{22}$ | 1 | 0 | 1 | 1 | 0 | 8 |
| $A_7$ | 0 | 0 | 1 | 1 | 1 | 4 | $A_{23}$ | 1 | 0 | 1 | 1 | 1 | 9 |
| $A_8$ | 0 | 1 | 0 | 0 | 0 | 3 | $A_{24}$ | 1 | 1 | 0 | 0 | 0 | 8 |
| $A_9$ | 0 | 1 | 0 | 0 | 1 | 4 | $A_{25}$ | 1 | 1 | 0 | 0 | 1 | 9 |
| $A_{10}$ | 0 | 1 | 0 | 1 | 0 | 4 | $A_{26}$ | 1 | 1 | 0 | 1 | 0 | 9 |
| $A_{11}$ | 0 | 1 | 0 | 1 | 1 | 5 | $A_{27}$ | 1 | 1 | 0 | 1 | 1 | 10 |
| $A_{12}$ | 0 | 1 | 1 | 0 | 0 | 5 | $A_{28}$ | 1 | 1 | 1 | 0 | 0 | 10 |
| $A_{13}$ | 0 | 1 | 1 | 0 | 1 | 6 | $A_{29}$ | 1 | 1 | 1 | 0 | 1 | 11 |
| $A_{14}$ | 0 | 1 | 1 | 1 | 0 | 6 | $A_{30}$ | 1 | 1 | 1 | 1 | 0 | 11 |
| $A_{15}$ | 0 | 1 | 1 | 1 | 1 | 7 | $A_{31}$ | 1 | 1 | 1 | 1 | 1 | 12 |

Here *FR* means the Fibonacci representations and *W* the weights of digits.

The analysis of Table 4.1 allows finding the following peculiarities of the Fibonacci representations of positive integers in the Fibonacci 1-code (4.4). By using the 5-digit Fibonacci 1-code (4.4), we can represent 13 positive integers in the range from 0 to 12, inclusively. Note that the number 13 is the Fibonacci number with the index 7, i.e. $F_1(7) = F_7 = 13$. The result of Table 4.1 is partial case of the following general theorem [14].

**Theorem 4.1.** *For the given integers $n \geq 0$ and $p \geq 0$ by using the n-digit Fibonacci p-code we can represent $F_p(n+p+1)$ positive integers in the range from 0 to $F_p(n+p) - 1$, inclusively.*

Note that for the case $p = 0$ Theorem 4.1 is reduced to the well-known theorem about the range of number representation, equal to $2^n$ for the $n$-digit binary code (4.3).

### 4.1.4. *Multiplicity of number representations*

A multiplicity of number representations (Fibonacci representations) in the form (4.5) is the most important peculiarity of Fibonacci $p$-codes (4.1) for the case $p > 0$ in comparison to the binary code (4.3). With the exception of the minimal number

$$N_{\min} = 0 = \underbrace{00...0}_{n} \qquad (4.7)$$

and the maximal number

$$N_{\max} = F_p(n+p) - 1 = \underbrace{11...1}_{n}, \qquad (4.8)$$

the other positive integers from the range $[0, F_p(n+p) - 1]$ have more than one Fibonacci representations in the form (4.5).

Let us consider now the mapping of positive integers onto the 5-digit Fibonacci representations in accordance with Table 4.2 ($p = 1$).

Table 4.2. Mapping of positive integers onto Fibonacci 1-representations.

| |
|---|
| $0 = \{A_0\}$ |
| $1 = \{A_1, A_2\}$ |
| $2 = \{A_3, A_4\}$ |
| $3 = \{A_5, A_6, A_8\}$ |
| $4 = \{A_7, A_9, A_{10}\}$ |
| $5 = \{A_{11}, A_{12}, A_{16}\}$ |
| $6 = \{A_{13}, A_{14}, A_{17}, A_{18}\}$ |
| $7 = \{A_{15}, A_{19}, A_{20}\}$ |
| $8 = \{A_{21}, A_{22}, A_{24}\}$ |
| $9 = \{A_{23}, A_{25}, A_{26}\}$ |
| $10 = \{A_{27}, A_{28}\}$ |
| $11 = \{A_{29}, A_{30}\}$ |
| $12 = \{A_{31}\}$ |

Here, $\{A_0, A_1, A_2, \ldots, A_{31}\}$ are the binary code combinations, taken from Table 4.1.

## 4.2. The minimal form and redundancy of Fibonacci *p*-codes

### 4.2.1. *Zeckendorf's sums*

Many number theorists know about *Zeckendorf sums*, but know a little about the man, who is the author of these sums. We can get the following biographical data about Edouard Zeckendorf from Wikipedia biographical article [80]:

*"Edouard Zeckendorf (2 May 1901–16 May 1983) was a Belgian doctor, army officer and mathematician. In mathematics, he is best known for his work on Fibonacci numbers and in particular for proving Zeckendorf's theorem. Zeckendorf was born in Liège in 1901. He was the son of a Dutch dentist. In 1925, Zeckendorf graduated as a medical doctor from the University of Liège and joined the Belgian Army medical corps. When Germany invaded Belgium in 1940, Zeckendorf was taken prisoner and remained a prisoner of war until 1945."*

Figure 4.1. Edouard Zeckendorf (1901–1983).

Fibonacci-Association recognizes Edouard Zeckendorf as one of the famous Fibonacci-mathematicians of the 20th century, though his basic education did not include mathematics. Mathematics was Zeckendorf's hobby. Through the years, Zeckendorf published several mathematical articles. The paper on *Zeckendorf's sums*, published in 1939, was the most important of them. In this paper, Zeckendorf proved that each positive integer can be represented as a unique sum of the non-adjacent Fibonacci numbers, as exemplified below:

$$38 = 34 + 3 + 1;\ 39 = 34 + 5;\ 40 = 34 + 5 + 1;$$
$$41 = 34 + 5 + 2;\ 42 = 34 + 8. \tag{4.9}$$

Note that *Zeckendorf's sums* (4.9) has direct relation to such important notion of Fibonacci *p*-codes (4.1) as the *minimal form* (see below).

### 4.2.2. *"Convolution" and "devolution" for the Fibonacci 1-code*

In Section 3.2 we have introduced the concepts of *convolution* (011→100) and *devolution* (100→011) for the "golden" representations (3.10), applied to the Φ-code (3.9) $N = \sum_i a_i \Phi^i$.

Note that Fibonacci 1-code (4.4) is discrete analog of the Φ-code (3.9) $N = \sum_i a_i \Phi^i$ and the concepts of *convolution* (011→100) and *devolution* (100→011) can be applied to Fibonacci representations:

(a) *Convolution*

$$7 = \begin{cases} 0 & 1 & 1 & 1 & 1 \\ 1 & 0 & 0 & 1 & 1 \\ 1 & 0 & 1 & 0 & 0 \end{cases} \tag{4.10}$$

(b) *Devolution*

$$5 = \begin{cases} 1 & 0 & 0 & 0 & 0 \\ 0 & 1 & 1 & 0 & 0. \\ 0 & 1 & 0 & 1 & 1 \end{cases} \tag{4.11}$$

The *convolution* result 10100 in (4.10) is named "*convolute*" *Fibonacci representation* and the *devolution* result 01011 in (4.11) is named "*devolute*" *Fibonacci representation*. For $p = 1$ the "*convolute*" and "*devolute*" Fibonacci representations of the positive integer $N$ have peculiar indications. In particular, *in the "convolute" Fibonacci representations two bits of 1 together do not meet* and *in the "devolute" Fibonacci representations two bits of 0 together do not meet, starting from the highest bit of 1 of the Fibonacci representation (4.5).*

Consider now peculiarities of the *convolution* and *devolution* for the lowest digits of the Fibonacci representation (4.5). As it is well-known, for the case $p = 1$ the weights of the two lowest digits of the Fibonacci 1-code (4.4) is equal to 1 identically, that is, $F_1 = F_2 = 1$. And then the operations of the *devolution* and *convolution* for these digits are performed as follows:

10→01 (*devolution*) and 01→10 (*convolution*).

### 4.2.3. *The base of the Fibonacci p-code*

For the case $p = 0$ the base of the binary system (4.3) is calculated as the ratio of the adjacent digit weights, that is,

$$\frac{2^k}{2^{k-1}} = 2.$$

Apply this principle to the Fibonacci $p$-code (4.1) and consider the ratio

$$\frac{F_p(k)}{F_p(k-1)}. \tag{4.12}$$

A limit of the ratio (4.12) for $k \to \infty$ is the *base of the Fibonacci p-code* (4.1). As it is shown above (Chapter 1), the limit of (4.12) is equal:

$$\lim_{k\to\infty} \frac{F_p(k)}{F_p(k-1)} = \Phi_p, \tag{4.13}$$

where $\Phi_p$ is the golden $p$-proportion.

This means that the base of the Fibonacci $p$-code (4.1) for the case $p > 0$ is the irrational number $\Phi_p$ and hence Fibonacci $p$-codes (4.1) are a new class of positional numeral systems with irrational bases.

### 4.2.4. *The minimal form of the Fibonacci p-code*

The following theorem is of great significance for the theory of the Fibonacci $p$-codes [14].

**Theorem 4.2.** *For the given integers $p \geq 0$ and $n \geq p + 1$ the arbitrary positive integer $N$ can be represented in the following unique form:*

$$N = F_p(n) + R_1 \tag{4.14}$$

where

$$0 \leq R_1 < F_p(n-p). \tag{4.15}$$

For the case $p = 1$ the formulas (4.14), (4.15) are reduced to the following:

$$N = F_n + R_1,\ 0 \leq R_1 < F_{n-1}. \tag{4.16}$$

Note that for the case $p = 0$ we have: $F_0(n) = 2^{n-1}$ and then the formulas (4.14) and (4.15) take the following well-known (for the "binary" arithmetic) formulas:

$$N = 2^{n-1} + R_1,\ 0 \leq R_1 < 2^{n-1}.$$

If we represent the integer $N$ according to the formulas (4.14), (4.15) and then all the remainders $R_1$, $R_2$, ..., $R_k$, arising as a result of this representation, according to the same formulas (4.14), (4.15) up to obtaining the remainder, equal to 0, we get a peculiar representation of the positive integer $N$ in the Fibonacci 1-code (4.4). Its peculiarity consists in the fact that in the Fibonacci representation of the positive integer $N$, given by (4.4), no less than one bit of 0 follow after every bit $a_l = 1$ from the left to the right.

Such Fibonacci representation of the positive integer $N$ is called *minimal form of the positive integer N in the Fibonacci* 1*-code*. This name reflects the fact that for the case $p = 1$ the *minimal form* of the positive integer $N$ has a minimal number of bits of 1 in the Fibonacci representation of the Fibonacci 1-code (4.4) among all Fibonacci representations of the same positive integer $N$.

For example, by using the above algorithm, we can obtain the following *minimal form* of the number 25 in the Fibonacci 1- code (Table 4.3).

Table 4.3. Example of the minimal form of the Fibonacci 1- code.

| $F_i$ | 55 | 34 | 21 | 13 | 8 | 5 | 3 | 2 | 1 | 1 |
|---|---|---|---|---|---|---|---|---|---|---|
| 25 = | 0 | 0 | **1** | **0** | 0 | 0 | **1** | **0** | **1** | **0** |

The *minimal form* of the number 25 has the following algebraic interpretation:

$$25 = 21 + 3 + 1. \tag{4.17}$$

Note that the sum (4.17) is the example of *Zeckendorf's sum* for the number 25 similar to the *Zeckendorf's sums* (4.9).

A peculiarity of the *minimal form* of the number 25, given in Table 4.3, consists of the following. The bits of 0 follow obligatory after every bit of 1 from the left to the right in the Fibonacci representation of the number 25 (see the binary combinations **10** in bold). This is the main property of the *minimal form.*

**Corollary from Theorem 4.2.** For the given $p$ ($p = 0, 1, 2, 3, \ldots$) every positive integer $N$ has the only *minimal form* in the Fibonacci $p$-code (4.1).

This means that there is a one-to-one mapping of the set of positive integers onto the set of the *minimal forms* of the Fibonacci $p$-code (4.1).

In the book [14] the following theorem is proved.

**Theorem 4.3.** *For a given integer $p \geq 0$ by using the n-digit Fibonacci 1-code we can represent in the minimal form $F_{n+1}$ positive integers in the range from 0 to $F_{n+1}-1$, inclusively.*

For the case of Fibonacci 1-code (4.4) the *minimal form* has very simple indication: *in the minimal form two bits of 1 together do not meet.* But the "*convolute*" *Fibonacci representation*, considered above, has the same property. This means that for the Fibonacci 1-code (4.4) the *"convolute" Fibonacci representation* coincides with the *minimal form* and the reduction of the Fibonacci representation to its *minimal form* can be performed by using *convolutions*. The example (4.10) demonstrates the process of the reduction of the initial Fibonacci representation 01111 to its *minimal form* (01111→10011→10100).

### 4.2.5. *Redundancy of the Fibonacci 1-code*

For $p = 0$, the Fibonacci 0-code (classic binary code) is non-redundant. This means that all the binary words of the binary code (4.3) are *allowed*, i.e. the binary code (4.3) is of *"zero" redundancy* and cannot be used for error detection and correction in computers and their structures. For $p > 0$ all the Fibonacci $p$-codes (4.2) are redundant and this important property can be used for error detection. For the example (4.10), the redundancy of the Fibonacci 1-code (4.4) shows itself in *multiplicity* of the Fibonacci representations of one and the same positive integer $N = 7$ and in the existence of the *minimal form* 10100.

As it is shown above, the amount of the Fibonacci $p$-codes (4.1) is theoretically infinite. The question is: which of the Fibonacci $p$-codes (4.1) are the most suitable for practical implementation? Studies have shown [14] that the Fibonacci 1-code (4.4) is the simplest in terms of technical implementation. That is why, the main attention in this chapter is paid to the study of Fibonacci 1-code (4.4) and its technical applications.

Theorem 4.3 allows calculating the relative code redundancy of the Fibonacci 1-code (4.4) in comparison with the classic binary code (4.3).

We will calculate the relative code redundancy $R$ by the following well-known formula:

$$R = \frac{n-m}{n} = 1 - \frac{m}{n}, \tag{4.18}$$

where $n$ and $m$ are the number of digits of the Fibonacci 1-code (4.4) and the binary code (4.3), respectively, for the representation of one and the same range of numbers.

Note that the relative code redundancy, which is given by (4.18), characterizes a relative increasing of the number of digits in the

Fibonacci 1-code (4.4) in the comparison to the binary code (4.3) for the representation of one and the same range of numbers.

It follows from Theorem 4.3 that we can represent the $F_{n+1}$ positive integers in the $n$-digit Fibonacci 1-code (4.4), if we use only the *minimal forms* for the representation of numbers. To represent this range of numbers in the binary system (4.3), we need approximately

$$m \approx \log_2 F_{n+1} \tag{4.19}$$

digits.

By substituting (4.19) into (4.18), we get the following formula for the calculation of the relative code redundancy of the Fibonacci 1-code (4.4):

$$R = 1 - \frac{m}{n} \approx 1 - \frac{\log_2 F_{n+1}}{n}. \tag{4.20}$$

We can express the Fibonacci number $F_{n+1}$ through the golden ratio $\Phi \approx 1.618$ by using Binet's formula (1.50) $F_n = \begin{cases} \dfrac{\Phi^n + \Phi^{-n}}{\sqrt{5}} \text{ for } n = 2k+1 \\ \dfrac{\Phi^n - \Phi^{-n}}{\sqrt{5}} \text{ for } n = 2k \end{cases}.$

For the case $n + 1$ and $n$ large we can write Binet's formula (1.50) in the following approximate form:

$$F_{n+1} \approx \frac{\Phi^{n+1}}{\sqrt{5}}. \tag{4.21}$$

By substituting the approximate value $F_{n+1}$, given by (4.21), into formula (4.20), we get the following approximate formula:

$$R \approx 1 - \frac{\log_2\left(\dfrac{\Phi^{n+1}}{\sqrt{5}}\right)}{n} = 1 - \frac{n\log_2\Phi + \log_2\Phi - \log_2\sqrt{5}}{n}. \tag{4.22}$$

For the case $n \to \infty$, the formula (4.22) is simplified and takes the following form:

$$R \approx 1 - \log_2 \Phi = 0.306\ (30.6\%). \tag{4.23}$$

Thus, the limiting value of the relative code redundancy of the Fibonacci 1-code (4.4) is a constant value, equal to 0.306 (30.6%).

### 4.2.6. *Surprising analogies between the Fibonacci 1-code and genetic code*

Among the biological concepts, having a level of general scientific significance and well formalized, the genetic code occupies a special place. A discovery of the well-known fact of striking simplicity of the basic principles of the genetic code falls into the major modern discoveries of human science. This simplicity is due to the fact that the inheritable information is encoded by the texts of the three-alphabetic words — *triplets* or *codonums*, compounded on the basis of the alphabet, consisting of four characters, which are nitrogen bases: ***A*** (adenine), ***C*** (cytosine), ***G*** (guanine), ***T*** (thiamine). The given recording system is unique for all boundless set of miscellaneous alive organisms and is called *genetic code* [81].

It is known [81] that by using three-alphabetic *triplets* or *codonums*, we can encode the 21 items, which include 20 amino acids and one additional item called *stop-codonum* (sign of the punctuation). It is clear that $4^3 = 64$ different combinations (from four by three nitrogen bases) are used for encoding 21 items. In this connection some of the 21 items are encoding by several triplets. It is called *degeneracy of genetic code*. Finding conformity between triplets and amino acids (or signs of the punctuation) is named the *decryption of genetic code*.

Consider now the 6-digit Fibonacci 1-code (4.4), which uses 6 Fibonacci numbers 1, 1, 2, 3, 5, 8 as digit weights:

$$N = a_6 \times 8 + a_5 \times 5 + a_4 \times 3 + a_3 \times 2 + a_2 \times 1 + a_1 \times 1 \tag{4.24}$$

The following are the surprising analogies between the 6-digit Fibonacci 1-code (4.24) and *genetic code*:

(1) *The first analogy*. For the representation of numbers, the 6-digit binary Fibonacci code uses $2^6 = 64$ binary combinations from 000000 up to 111111 which coincides with the number of the triplets of the genetic code ($4^3 = 64$).

(2) *The second analogy*. By using the 6-digit Fibonacci 1-code (4.24), we can represent 21 integers, by starting from the minimal number 0, encoded by the 6-digit binary combination 00000, and by ending with the maximal number 20, encoded by the 6-digit binary combination 111111. Note, by using triplet's coding we can also represent 21 items including 20 amino acids and one additional object, the *stop-codonum*, which means the *sign of the punctuation* indicates a termination of protein synthesis.

(3) *The third analogy*. The main feature of the Fibonacci 1-code (4.24) is a *multiplicity* of number representation. With the exception for the minimal number 0 and the maximal number 20, which have the only code representations 000000 and 111111, respectively, all the other numbers from 1 up to 19 have multiple representations in the 6-digit Fibonacci 1-code (4.24), that is, they use not less than two code combinations for their representation. It is necessary to note that the genetic code has similar property called the *degeneracy* of the genetic code.

Thus, between the Fibonacci 1-code (4.24) and the genetic code, based on triplet's representation of amino acids, there are very interesting analogies, which allow one to consider the Fibonacci 1-code (4.4) as a peculiar class of the redundant codes among other ways of redundant coding. One may assume that a study of the Fibonacci 1-code (4.4) can have great interest to computer science. It is possible to assume from the above analogies that the Fibonacci 1-code (4.4) can become rather useful at designing the bio-computers, based on DNA.

## 4.3. Fibonacci arithmetic

### 4.3.1. *Comparison of numbers in the Fibonacci 1-code*

Let us deduce now a rule of the number comparison in the Fibonacci 1-code (4.4). We begin from simple example. Let us compare two positive integers $A$ and $B$ (see Table 4.4), which are represented by the 8-digit *minimal forms* of the Fibonacci 1-code (4.4).

Table 4.4. *Minimal forms A and B.*

| $i$ | 8 | 7 | 6 | 5 | 4 | 3 | 2 | 1 |
|---|---|---|---|---|---|---|---|---|
| $W_i$ | 21 | 13 | 8 | 5 | 3 | 2 | 1 | 1 |
| $A$ | 1 | 0 | $a_6$ | $a_5$ | $a_4$ | $a_3$ | $a_2$ | $a_1$ |
| $B$ | 0 | 1 | 0 | $b_5$ | $b_4$ | $b_3$ | $b_2$ | $b_1$ |

Here $i \in \{1,2,3,4,5,6,7,8\}$ is the number of $i$th digit and $W_i = F_i$ ($i$ = 1, 2, 3, …, 8) is the weight of $i$th digit of the 8-digit Fibonacci 1-code (4.4). The *minimal form* of the number $A$ begins from the bit $a_8 = 1$ and $a_7 = 0$, according to the property of the *minimal form*.. The *minimal forms* of the number $B$ begins from the bits $b_8 = 0$, $b_7 = 1$ and $b_6 = 0$, according to the property of the *minimal forms*. The other bits $a_6a_5a_4a_3a_2a_1$ of the number $A$ and the other bits $b_5b_4b_3b_2b_1$ of number $B$ are the *minimal forms*. The *minimal forms* of the numbers $A$ and $B$ (Table 4.4) have the following algebraic interpretations:

$$A = 1\times 21 + 0\times 13 + a_6 \times 8 + a_5 \times 5 + a_4 \times 3 + a_3 \times 2 + a_2 \times 1 + a_1 \times 1 \quad (4.25)$$

$$B = 0\times 21 + 1\times 13 + b_6 \times 8 + b_5 \times 5 + b_4 \times 3 + b_3 \times 2 + b_2 \times 1 + b_1 \times 1. \quad (4.26)$$

Let us now compare $A$ and $B$, represented by the sums (4.25), (4.26). To do this, we calculate the minimal possible value of $A$ with $a_8 = 1$ and the maximal possible value of $B$ with $b_8 = 0$, $b_7 = 1$ and $b_6 = 0$. Recall that the Fibonacci representations of $A$ and $B$ are *minimal forms*. It is clear that the minimal value of number $A$, given by the sum (4.25), is equal $A_{\min} = 21$, when all the other bits $a_6 = a_5 = a_4 = a_3 = a_2 = a_1 = 0$. On the other hand, the maximal value of $B$, given by the sum (4.26), according to the property of the *minimal form*, is equal to:

$$B_{\max} = 01010100$$

$$= 0\times 21 + 1\times 13 + 0\times 8 + 1\times 5 + 0\times 3 + 1\times 2 + 0\times 1 + 0\times 1 = 20. \quad (4.27)$$

It is clear that the comparison of the minimal value of $A$ ($A_{\min} = 21$) and the maximal value of $B$ ($B_{\max} = 20$) leads us to the following result: $A_{\min} > B_{\max}$.

From this consideration, it follows the simple rule of the comparison of two positive integers $A$ and $B$, represented in the *minimal form* of the Fibonacci 1-code (4.4). The comparison of two numbers $A$ and $B$, represented in the *minimal form*, is fulfilled from the left to the right (digit-by-digit), starting with the highest digit, until finding the first pair of the non-coincident digits of the comparable *minimal forms.* If the number $A$ has the bit of 1 and the number $B$ has the bit of 0 in the first pair of the non-coincident digits, then we can conclude that $A > B$. In the opposite case, we have: $A < B$ or $A = B$. If the *minimal forms* of the comparable numbers coincide for all digits, the numbers are equal between themselves, that is, $A = B$.

Note that *the comparison of numbers in the Fibonacci 1-code (4.4) is fulfilled similarly to the classic binary code (4.3), if the comparable numbers are represented in the minimal form.* This property (simplicity of number comparison) is one of the most important arithmetical advantages of the Fibonacci 1-code (4.4).

In [14] it is proved that this rule can be applied to all the Fibonacci $p$-codes (4.2).

### 4.3.2. *The basic micro operations*

As we mentioned above, the main distinction of the Fibonacci 1-code (4.4) from the binary code (4.3) is a *multiplicity* of Fibonacci representations of one and the same positive integer. By using the above micro-operations of *devolution* (011→100) and *convolution* (100→011), we can change the forms of Fibonacci representations of one and the same positive integer. This means that the binary 1's in the Fibonacci representation (4.5) can move to the left or to the right along Fibonacci representation (4.5) of the same number by using the micro-operations *devolution* (011→100) and *convolution* (100→011). Recall again that the fulfilment of these micro-operations does not change the number itself, that is, we will get the different Fibonacci representations of the same number. This fact allows one to develop the original approach to the Fibonacci arithmetic, based on the so-called *basic micro-operations.*

Let us introduce the following four *basic micro-operations*, used to fulfill logical and arithmetical operations over binary words:

| *Convolution* | *Devolution* | *Replacement* | *Absorption* |
|---|---|---|---|
| $100 \leftarrow 011$ | $100 \rightarrow 011$ | $\begin{bmatrix} 1 & & 0 \\ \downarrow & = & \\ 0 & & 1 \end{bmatrix}$ | $\begin{bmatrix} 1 & & 0 \\ \updownarrow & = & \\ 1 & & 0 \end{bmatrix}$ |

. (4.28)

Note that the noise-immune Fibonacci arithmetic, based on the above micro-operations (4.28), is described for the first time in [23].

Note that the *convolutions* and *devolutions*, shown in (4.28), are the simple code transformations, which are performed over the adjacent three bits of the Fibonacci representation of one and the same number $N$ in the Fibonacci 1-code (4.4).

The micro-operation of *replacement*

$$\begin{bmatrix} 1 & & 0 \\ \downarrow & = & \\ 0 & & 1 \end{bmatrix}$$

is a two-placed micro-operation, which is fulfilled over the same digits of two registers, the top register $A$ and the lower register $B$. Consider now the case, when the register $A$ has 1 in the $k$th digit and the register $B$ has 0 in the same $k$th digit (the condition for the *replacement*). The micro-operation of the *replacement* consists in moving the bit 1 from the $k$th digit of the top register $A$ to the $k$th digit of the lower register $B$. Note that this operation can only be fulfilled, if the bits of the $k$th digits of the registers $A$ and $B$ are equal to 1 and 0, respectively.

The micro-operation of *absorption*

$$\begin{bmatrix} 1 & & 0 \\ \updownarrow & = & \\ 1 & & 0 \end{bmatrix}$$

is a two-placed micro-operation for the condition, when the bits of 1 are in the $k$th digits of the top register $A$ and the lower register $B$. This micro-operation consists in mutual annihilation of the bits of 1 in the top and lower registers $A$ and $B$. After fulfilment of the micro-operation of *absorption* the bits of 1 are replaced by the bits of 0.

It is necessary to pay attention to the "technical" peculiarity of the above "basic micro-operations". At the register interpretation of these micro-operations, each micro-operation may be fulfilled by means of the inversion of the flip-flops, involved into the micro-operation. This means that each micro-operation is reduced to the flip-flops' switching.

### 4.3.3. *Logic operations*

We demonstrate a possibility to fulfill the simplest logic operations by means of the above *basic micro-operations* (4.28).

Let us perform now all possible *replacements* from the top register $A$ to the lower register $B$:

$$\begin{array}{cccccccccc} A & = & 1 & 0 & 0 & 0 & 1 & 0 & 1 \\ & & \downarrow & & & & \downarrow & & \\ B & = & 0 & 1 & 0 & 1 & 0 & 0 & 1 \\ \hline A' & = & 0 & 0 & 0 & 0 & 0 & 0 & 1 \\ B' & = & 1 & 1 & 0 & 1 & 1 & 0 & 1 \end{array}$$

As a result of the *replacement*, we get the two new binary words $A'$ and $B'$. We can see that the binary word $A'$ is a logic *conjunction* ($\wedge$) of the initial binary words $A$ and $B$, that is,

$$A' = A \wedge B$$

and the binary word $B'$ is a logic *disjunction* ($\vee$) of the initial binary words $A$ and $B$, that is,

$$B' = A \vee B.$$

A logic operation of the *module 2 addition* is fulfilled by means of the simultaneous fulfillment of all the possible *replacements* and *absorptions*. For example:

$$\begin{array}{ccccccccccccc} A & = & 1 & 0 & 1 & 0 & 0 & 1 & 1 & 0 & 1 & & \\ & & \updownarrow & & \downarrow & & & \downarrow & \downarrow & & \updownarrow & & \\ B & = & 1 & 1 & 1 & 0 & 0 & 1 & 1 & 0 & 1 & & . \\ \hline A' & = & 0 & 0 & 0 & 0 & 0 & 0 & 0 & 0 & 0 & = & \text{const}\,0 \\ B' & = & 0 & 1 & 1 & 0 & 0 & 1 & 1 & 0 & 0 & = & A \oplus B \end{array}$$

We can see that the results of this code transformation are two new binary words $A' = \text{const } 0$ and $B' = A \oplus B$. It is clear that the binary word $A' = \text{const } 0$ plays a role of checking binary word for the *module 2 addition* which is important for computer applications.

A logic operation of the *code A inversion* is reduced to the fulfillment of the *absorptions* over the initial binary word $A$ and the special binary word $B = \text{const } 1$:

$$\begin{array}{ccccccccccccc} A & = & 1 & 0 & 1 & 0 & 0 & 1 & 1 & 0 & 1 & & \\ & & \updownarrow & & \updownarrow & & & \updownarrow & \updownarrow & & \updownarrow & & \\ B & = & 1 & 1 & 1 & 1 & 1 & 1 & 1 & 1 & 1 & = & \text{const}\,1. \\ \hline A' & = & 0 & 0 & 0 & 0 & 0 & 0 & 0 & 0 & 0 & = & \text{const}\,0 \\ B' & = & 0 & 1 & 0 & 1 & 1 & 0 & 0 & 1 & 0 & = & \overline{A} \end{array}$$

The binary word $A' = \text{const } 0$ plays the role of checking binary word for the *inversion* which is important for computer applications.

## 4.4. Counting and subtracting of the binary 1's

Let us demonstrate now a possibility to fulfill the simplest arithmetical operations by using the *basic micro-operations* (4.28). We start with the operations of *counting* and *subtracting* of the binary 1's.

### 4.4.1. *Algorithm of summing Fibonacci counter*

Counting of the binary 1's in the Fibonacci 1-code (the *summing* Fibonacci counter) is realized with the help of the *convolutions*. For example, the transition of the initial Fibonacci representation, which is the *minimal form* of the number 4

$$\begin{array}{ccccccc} & & 5 & 3 & 2 & 1 & 1 \\ 4 & = & 0 & 1 & 0 & 1 & 0 \end{array}$$

to the next Fibonacci representation of the number 5 = 4 + 1 in the *summing* Fibonacci counter is fulfilled in the following manner:

$$\begin{array}{ccccccccc} & & 5 & 3 & 2 & 1 & 1 & & \\ 4 & = & 0 & 1 & 0 & 1 & 0 & + & 1 \\ 5 & = & 0 & 1 & 0 & 1 & 1 & & . \\ 5 & = & 0 & 1 & 1 & 0 & 0 & & \\ 5 & = & 1 & 0 & 0 & 0 & 0 & & \end{array}$$

Here in the top row we see Fibonacci numbers 5, 3, 2, 1, 1, which are the digit weights of the 5-digit Fibonacci 1-code (4.4). The second row is the representation of the number 4 in the *minimal form* (4 = 01010). We can see that the binary 1 is added (+1) to the lower digit of the Fibonacci representation 01010. As a result, the Fibonacci representation 4 = 01010 is transformed into the Fibonacci representation of the next number 5 = 01011 (the third row). After that, the Fibonacci representation 5 = 01011 is reduced to the *minimal form*.

This code transformation is fulfilled in 2 steps with the help of the *convolutions*. The first step is to fulfill the *convolution* for the three lower digits of the Fibonacci representation 5 = 01011 (the third row). Due to this operation, the Fibonacci representation of number 5 = 01011 is transformed to another Fibonacci representation of the same number 5 = 01100 (the fourth row). Then, we can fulfill the *convolution* for the next group 011 of the Fibonacci representation 5 = 01100 (the fourth row). In the fifth row we can see the *minimal form* of number 5 = 10000. Then we can continue *counting* the bits of 1 as follows:

$$6 = 10000 + 1 = 10001 = 10010,$$

$$7 = 10010 + 1 = 10011 = 10100.$$

If we add the binary 1 to the lower digit of number 7 = 10100, we get:

$$10100 + 1 = 10101 = 10110 = 11000 = 00000.$$

This situation is well known in computer engineering under the name of the *overfilling* of the *summing counter*.

### 4.4.2. *Algorithm of the subtracting Fibonacci counter*

A subtraction of the bit of 1 (the *subtracting* Fibonacci counter) is fulfilled with the help of the *devolutions*. For this purpose, the initial Fibonacci combination, represented in the *minimal form*, is reduced to the "devolute" or *maximal form* and then the binary 1 is subtracted from the lower digit:

$$\begin{array}{ccccccccc} & & 5 & 3 & 2 & 1 & 1 & & \\ 5 & = & 1 & 0 & 0 & 0 & 0 & & \\ 5 & = & 0 & 1 & 1 & 0 & 0 & & . \\ 5 & = & 0 & 1 & 0 & 1 & 1 & - & 1 \\ 4 & = & 0 & 1 & 0 & 1 & 0 & & \end{array}$$

Here in the second row we see the representation of number 5 in the *minimal form* (5 = 10000). In the third row we fulfill the *devolution* for the three higher digits (10000→01100) of the initial Fibonacci representation of number 5 = 10000. In the fourth row we fulfill the *devolution* for the next three digits of the Fibonacci representation of number 5 = 01100 (01100→01011). As a result, we get the *maximal form* of number 5 = 01011 (the fourth row). Then, we subtract the binary 1 (−1) from the lower digit of the Fibonacci representation 5 = 01011. The result of this transformation 4 = 01010 is represented in the fifth row. After that we can continue *subtracting* the 1's as follows:

4 = 01010 = 01001 (*devolution*)

3 = 01001 – 1 = 01000 (*subtracting* 1)

3 = 01000 = 00110 = 00101 (*devolution*)

2 = 00101 – 1 = 00100 (*subtracting* 1)

2 = 00100 = 0011 (*devolution*)

1 = 00011 – 1 = 00010 (*subtracting* 1)

1 = 00010 = 0001 (*devolution*)

0 = 00001 – 1 = 00000 (*subtracting* 1).

## 4.5. Fibonacci summation and subtraction

### 4.5.1. *Fibonacci summation*

The idea of the summation of two numbers $A$ and $B$ by using the *basic micro-operations* consists of the following. We have to move all the binary 1's from the top register $A$ to the lower register $B$. For this purpose we use the micro-operations of *replacement, devolution* and *convolution*. The result is formed in the register $B$.

For example, let us summarize the following numbers $A_0$ = 010100100 and $B_0$ = 001010100 as follows.

*The first step* of the summation consists in the *replacement* of all possible bit's of 1 from the register $A$ to the register $B$:

$$\begin{array}{lcccccccccc} A_0 & = & 0 & 1 & 0 & 1 & 0 & 0 & 1 & 0 & 0 \\ & & & \downarrow & & \downarrow & & & & & \\ B_0 & = & 0 & 0 & 1 & 0 & 1 & 0 & 1 & 0 & 0. \\ \hline A_1 & = & 0 & 0 & 0 & 0 & 0 & 0 & 1 & 0 & 0 \\ B_1 & = & 0 & 1 & 1 & 1 & 1 & 0 & 1 & 0 & 0 \end{array}$$

For this we apply the micro-operation of *replacement* to all digits of the initial numbers $A$ and $B$. However, this can be fulfilled only for those digits, where the condition of *replacement* is satisfied.

*The second step* is the fulfillment of all the possible *devolutions* in the binary word $A_1$ and all the possible *convolutions* in the binary word $B_1$, that is,

$$\begin{array}{lcccccccccc} A_1 & = & 0 & 0 & 0 & 0 & 0 & 0 & 1 & 0 & 0 \\ B_1 & = & 0 & 1 & 1 & 1 & 1 & 0 & 1 & 0 & 0 \\ & & & & & \Downarrow & & & & & . \\ A_2 & = & 0 & 0 & 0 & 0 & 0 & 0 & 0 & 1 & 1 \\ B_2 & = & 1 & 0 & 0 & 1 & 1 & 0 & 1 & 0 & 0 \end{array}$$

*The third step* is the *replacement* of all the possible bits of 1 from register $A$ to register $B$:

$$\begin{array}{lcccccccccc} A_2 & = & 0 & 0 & 0 & 0 & 0 & 0 & 0 & 1 & 1 \\ & & & & & & & & & \downarrow & \downarrow \\ B_2 & = & 1 & 0 & 0 & 1 & 1 & 0 & 1 & 0 & 0. \\ \hline A_3 & = & 0 & 0 & 0 & 0 & 0 & 0 & 0 & 0 & 0 \\ B_3 & = & 1 & 0 & 0 & 1 & 1 & 0 & 1 & 1 & 1 \end{array}$$

The summation is over, because all bits of 1 have moved from register $A$ to register $B$. After reducing the binary word $B_3$ to the *minimal form*, we get the sum $B_3 = A_0 + B_0$, represented in the *minimal form*:

$$B_3 = 100110111 = 101001001 = 101001010 = A_0 + B_0.$$

Thus, the summation is reduced to a sequential fulfillment of the micro-operations of the *replacement* for the two binary words $A$ and $B$ and the micro-operations of the *convolution* for the binary word $B$ and the *devolution* for the binary word $A$.

### 4.5.2. *Fibonacci subtraction*

The idea of the Fibonacci subtraction of number $B$ from number $A$ by using the *basic micro-operations* consists in the mutual *absorptions* of the binary 1's in the Fibonacci representations of numbers $A$ and $B$, until one of them becomes equal to 0. To realize this idea we have to fulfill sequentially the mentioned micro-operations of *absorption* for the Fibonacci representations $A$ and $B$ and then the micro-operations of *devolution* for the Fibonacci representations $A$ and $B$. The subtraction result is always formed in the register of the bigger number. If the result is formed in the top register $A$, it follows that the sign of the subtraction result is "+", in the opposite case the subtraction result has the sign "–".

Let us demonstrate now this idea on the following example. Let us subtract the number $B_0 = 101010010$ from the number $A_0 = 101001000$, represented in the *minimal form* of the Fibonacci 1-code.

*The first step* is the *absorption* of all possible binary 1's in the initial Fibonacci representations $A_0$ and $B_0$:

$$\begin{array}{lcccccccccc}
A_0 & = & 1 & 0 & 1 & 0 & 0 & 1 & 0 & 0 & 0 \\
 & & \updownarrow & & \updownarrow & & & & & & \\
B_0 & = & 1 & 0 & 1 & 0 & 1 & 0 & 0 & 1 & 0. \\
\hline
A_1 & = & 0 & 0 & 0 & 0 & 0 & 1 & 0 & 0 & 0 \\
B_1 & = & 0 & 0 & 0 & 0 & 1 & 0 & 0 & 1 & 0
\end{array}$$

*The second step* is the *devolutions* for the Fibonacci representations $A_1$ and $B_1$:

$$\begin{array}{lcccccccccc}
A_1 & = & 0 & 0 & 0 & 0 & 0 & 1 & 0 & 0 & 0 \\
B_1 & = & 0 & 0 & 0 & 0 & 1 & 0 & 0 & 1 & 0 \\
 & & & & & \Downarrow & & & & & \\
A_2 & = & 0 & 0 & 0 & 0 & 0 & 0 & 1 & 1 & 0 \\
B_2 & = & 0 & 0 & 0 & 0 & 0 & 1 & 1 & 0 & 1
\end{array}$$

*The third step* is the *absorptions* for the Fibonacci representations $A_2$ and $B_2$:

$$\begin{array}{lcccccccccc}
A_2 & = & 0 & 0 & 0 & 0 & 0 & 0 & 1 & 1 & 0 \\
 & & & & & & & & \updownarrow & & \\
B_2 & = & 0 & 0 & 0 & 0 & 0 & 1 & 1 & 0 & 1. \\
\hline
A_3 & = & 0 & 0 & 0 & 0 & 0 & 0 & 0 & 1 & 0 \\
B_3 & = & 0 & 0 & 0 & 0 & 0 & 1 & 0 & 0 & 1
\end{array}$$

*The fourth step* is the *devolutions* for the Fibonacci representations $A_3$ and $B_3$:

$$\begin{array}{lcccccccccc}
A_3 & = & 0 & 0 & 0 & 0 & 0 & 0 & 0 & 1 & 0 \\
B_3 & = & 0 & 0 & 0 & 0 & 0 & 1 & 0 & 0 & 1 \\
 & & & & & \Downarrow & & & & & \\
A_4 & = & 0 & 0 & 0 & 0 & 0 & 0 & 0 & 0 & 1 \\
B_4 & = & 0 & 0 & 0 & 0 & 0 & 0 & 1 & 1 & 1
\end{array}$$

*The fifth step* is the *absorptions* for the Fibonacci representations $A_4$ and $B_4$:

$$
\begin{array}{lcccccccccc}
A_4 & = & 0 & 0 & 0 & 0 & 0 & 0 & 0 & 0 & 1 \\
 & & & & & & & & & & \updownarrow \\
B_4 & = & 0 & 0 & 0 & 0 & 0 & 0 & 1 & 1 & 1. \\
\hline
A_5 & = & 0 & 0 & 0 & 0 & 0 & 0 & 0 & 0 & 0 \\
B_5 & = & 0 & 0 & 0 & 0 & 0 & 0 & 1 & 1 & 0
\end{array}
$$

The subtraction is over because $A_5 = 000000000$. After reducing the Fibonacci representation $B_5$ to the *minimal form* we get the subtraction result:

$$B_5 = 000001000.$$

The subtraction result is in register $B$. This means that the sign of the subtraction result is "–", that is, the difference of the numbers $A - B$ is equal to:

$$D = A - B = -000001000.$$

If we code the sign "–" by the bit of 1, then we can represent the difference $D$ as follows:

$$D = A - B = 1.000001000.$$

## 4.6. Fibonacci multiplication and division

### 4.6.1. *The "binary" multiplication*

To find the algorithms of the Fibonacci multiplication and division we will use an analogy to the classic binary multiplication and division. We start from the multiplication. To multiply two numbers $A$ and $B$ in the classic binary code (4.3), that is, to get the product $P = A \times B$, we should represent the multiplier $B$ in the form of the $n$-digit binary code (4.3). Then, the product $P = A \times B$ can be written in the following form:

$$
\begin{aligned}
P = A \times B = A \times b_n 2^{n-1} + A \times b_{n-1} 2^{n-2} + \ldots \\
+A \times b_i 2^{i-1} + \ldots + A \times b_1 2^0,
\end{aligned}
\tag{4.29}
$$

where $b_i \in \{0,1\}$ is the binary numerals of the multiplier $B$. It follows from (4.29) that the binary multiplication is reduced to forming the partial products of the kind $A \times b_i 2^{i-1}$ and their summation. The partial product $A \times b_i 2^{i-1}$ is formed by shifting the Fibonacci representation of number $A$ to the left in the $(i-1)$ digits.

The binary multiplication algorithm, based on (4.29), has a long history and goes back to the Egyptian *doubling method*, described in Chapter 2.

### 4.6.2. *Fibonacci multiplication*

The analysis of the Egyptian *doubling method* (see Chapter 2) allows the following method of the *Fibonacci multiplication* for the general case of $p$.

Let us consider now the product $P = A \times B$, where the numbers $A$ and $B$ are represented in the Fibonacci $p$-code (4.2). By using the representation of the multiplier $B$ in the Fibonacci $p$-code (4.2), we can represent the product $P = A \times B$ as follows:

$$P = A \times B = A \times b_n F_p(n) + A \times b_{n-1} F_p(n-1) + \ldots$$
$$+ A \times b_i F_p(i) + \ldots + A \times b_1 F_p(1), \tag{4.30}$$

where $F_p(i)$ $(i = 1, 2, 3, \ldots, n)$ are the Fibonacci $p$-numbers.

Note that the sum (4.30) is a generalization of the sum (4.29), which underlies the algorithm of the "binary" multiplication. The algorithm of the Fibonacci multiplication follows from the sum (4.30). The multiplication is reduced to the summation of the partial products of the kind $A \times b_i F_p(i)$. They are formed from the multiplier $A$ according to the special procedure, which is an analog of the Egyptian multiplication. Demonstrate now the *Fibonacci multiplication* for the case of the simplest Fibonacci 1-code (4.4).

**Example 4.1.** Find the following product: $41 \times 305$.
**Solution** is in Table 4.5.

Table 4.5. Example of Fibonacci multiplication.

| $F$ | $G$ | $P$ |
|---|---|---|
| 1 | 305 | |
| 1 | 305 | |
| **/2** | **610** | **→610** |
| 3 | 915 | |
| **/5** | **1525** | **→1525** |
| 8 | 2440 | |
| 13 | 3965 | |
| 21 | 6405 | |
| **/34** | **10370** | **→10370** |
| $\mathbf{41 = 34 + 5 + 2}$ | $\mathbf{41 \times 305}$ | $\mathbf{= 12505}$ |

Let us explain Table 4.5:

1. Construct Table 4.5, consisting of three columns: $F$, $G$ and $P$.
2. Insert the Fibonacci numbers 1, 1, 2, 3, 5, 8, 13, 21, 34 into the $F$-column of Table 4.5.
3. Insert the generalized Fibonacci 1-sequence: 305, 305, 610, …, 10370, which is formed from the first multiplier 305 according to the "Fibonacci recurrent relation" $G_i = G_{i-1} + G_{i-2}$, to the $G$-column.
4. Mark by the inclined line (/) all the $F$-numbers that give the second multiplier in the sum (**41 = 34 + 5 + 2**).
5. Mark in black font all the $G$-numbers **610**, **1525**, **10370**, corresponding to the marked $F$-numbers and rewrite them to the $P$-column.
6. By summarizing all the $P$-numbers **610 + 1525 + 10370**, we get the product: **41 × 305 = 12505**.

This multiplication algorithm is easily generalized for the case of the Fibonacci $p$-codes (4.2).

### 4.6.3. *Fibonacci division*

We can apply the above Egyptian method of division to construct the algorithm of the *Fibonacci division.* Let us consider this method for the following example.

**Example 4.2.** Let us divide the number **481** (the dividend) by the number **13** (the divisor) in the Fibonacci 1-code.
**Solution:**

***The first stage***

1. Construct the table, consisting of three columns: *F*, *G* and *D* (see Table 4.6).

Table 4.6. The first stage of the Fibonacci division.

| $F$ | $G$ | $D$ |
|---|---|---|
| 1 | 13 | $\leq 481$ |
| 1 | 13 | $\leq 481$ |
| 2 | 26 | $\leq 481$ |
| 3 | 39 | $\leq 481$ |
| 5 | 65 | $\leq 481$ |
| 8 | 104 | $\leq 481$ |
| 13 | 169 | $\leq 481$ |
| 21 | 273 | $\leq 481$ |
| **/34** | **442** | $\leq 481$ |
| 55 | 715 | $> 481$ |
| $R_1 =$ | **481 − 442** | **= 39** |

Let us explain Table 4.6:

2. Insert the Fibonacci numbers 1, 1, 2, 3, 5, 8, 13, 21, 34, 55 into the *F*-column of Table 4.6.
3. Insert the generalized Fibonacci 1-sequence: 13, 13, 26, 39, …, 615, formed from the divisor 13 according to the "Fibonacci recurrent relation" $G_i = G_{i-1} + G_{i-2}$, to the *G*-column.

4. Compare sequentially every *G*-number with the dividend **481**, inscribed into the *D*-column, and fix the results of comparison (≤ or >) until we obtain the first comparison result of the kind (>): **715 > 481**.
5. Mark by the incline line (/) the *F*-number of **34**, corresponding to the preceding *G*-number of **442**.
6. Calculate the difference: $\boldsymbol{R_1}$ **= 481 – 442 = 39**.

***The second stage***

The second stage of the Fibonacci division (Table 4.7) is the repetition of the first stage but we use instead of **481** the difference $\boldsymbol{R_1}$ = **39**, taken from Table 4.6.

Table 4.7. The second stage of the Fibonacci division.

| *F* | *G* | *D* |
|---|---|---|
| 1 | 13 | ≤39 |
| 1 | 13 | ≤39 |
| 2 | 26 | ≤39 |
| **/3** | **39** | ≤39 |
| 5 | 65 | >39 |
| $\boldsymbol{R_2}$ = | **39 – 39** | **= 0** |

Because the second difference $\boldsymbol{R_2}$ = **39 – 39 = 0**, this means the Fibonacci division is over. The result of the division is equal to the sum of all the marked *F*-numbers obtained on all stages and taken from Table 4.6 (the number **34**) and Table 4.7 (the number **3**), that is: **34 + 3 = 37**.

## 4.7. Useful applications of the Fibonacci code

### 4.7.1. *Error detection*

As it is well known, all errors, arising in functional devices of computer, can be divided into two groups: 1) the "*soft errors*," which result from the *random effects* on electronic elements, and the "*hard errors*," which result from the *constant failures* of electronic elements. Both types of errors are dangerous and may lead to "false" data on the output of computer.

As for the "*hard errors*," they can be detected by the register for the reduction of the Fibonacci code to the *minimal form* (see below). This register is an important device of all arithmetic units and thanks to this device all the Fibonacci arithmetical devices become self-checking devices. This is the first important advantage of the Fibonacci code.

### 4.7.2. *Potential error-detection ability of the Fibonacci 1-code*

Let us consider now the error detection in such an important computer unit as an electronic memory. Figure 4.3 shows the principle of error detection for the Fibonacci 1-code.

$$\{F_{n+1}\} \Rightarrow\Rightarrow\Rightarrow\Rightarrow\Rightarrow\Rightarrow\Rightarrow \{2^n\}$$

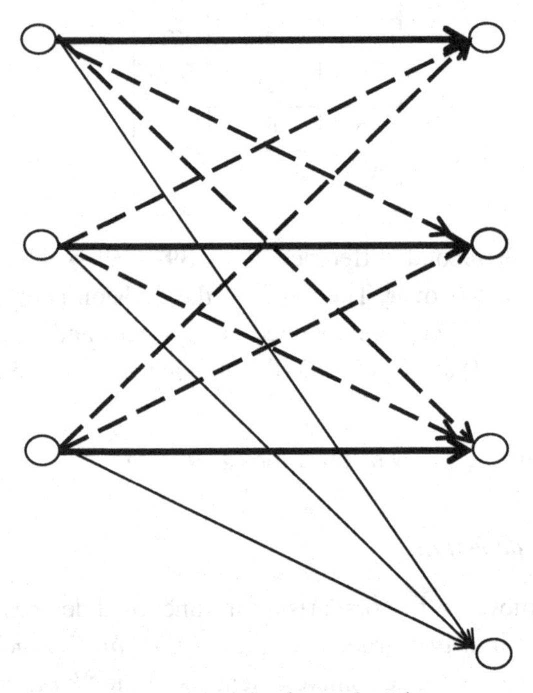

Fig. 4.3. Principle of error detection.

Figure 4.3 demonstrates the transitions of the $F_{n+1}$ *allowed* Fibonacci representations (*minimal forms*) into the $2^n$ binary code combinations, which include the $n$-bit *allowed* ($N_1 = F_{n+1}$) and $n$-bit *forbidden* ($N_2 = 2^n - F_{n+1}$) binary code combinations.

The set of all transitions in Fig. 4.3 are divided into three non-overlapping subsets: 1) *unmistakable* transitions (thick lines in Fig. 4.3); 2) *detectable erroneous* transitions (dash lines); 3) *undetectable erroneous* transitions (thin lines).

The *unmistakable* transitions correspond to the *true* transitions, when the *allowed* Fibonacci representations (*minimal forms*) pass into the same *minimal forms*. The example of the *unmistakable* transition is the following: $100100010 \to 100100010$. It is clear that the number of the *unmistakable* transitions $N_1$ is equal to the number of the *minimal forms* $F_{n+1}$, that is, $N_1 = F_{n+1}$.

The *undetectable erroneous* transition is such *erroneous* transition, when some *minimal form* passes into other *minimal form*. The example of the *undetectable erroneous* transition is the following:

$$\begin{array}{c} 100100010 \\ \Downarrow \\ 10\boxed{10}0\boxed{10}00. \end{array} \tag{4.31}$$

In the example (4.31), the distortion of the 4 bits in the initial *minimal form* 100100010 leads to another *minimal form* $10\boxed{10}0\boxed{10}00$; such error cannot be detected.

The error is detected, when the *allowed* Fibonacci representation (*minimal form*) passes into *forbidden* codeword. Such transitions are called *detectable erroneous* transitions. The example of the *detectable erroneous* transition is the following:

$$\begin{array}{c} 100100010 \\ \Downarrow \\ 1\boxed{1}0100\boxed{1}10. \end{array} \tag{4.32}$$

In the example (4.32), the distortion of the 2 bits in the initial *minimal form* 100100010 (the addition of 2 bits of 1) leads to the violation of the *minimal form* $1\boxed{1}0100\boxed{1}10$; such error is detected.

The potential error-detection ability of the $n$-bit Fibonacci 1-code (4.4) is determined by the relationship between the *allowed* and *forbidden* transitions.

It is clear that the total number of all the possible transitions for the diagram in Fig. 4.3 is equal to the product $P_1 = F_{n+1} \times 2^n$ and the number of the *detectable* transitions is equal to the product $P_2 = F_{n+1} \times (2^n - F_{n+1})$.

Then the *potential error-detection ability coefficient* $S_d$ for the $n$-bit Fibonacci 1-code (4.4) is determined as follows:

$$S_d = \frac{P_2}{P_1} = \frac{F_{n+1}\left(2^n - F_{n+1}\right)}{F_{n+1} \times 2^n} = 1 - \frac{F_{n+1}}{2^n}. \tag{4.33}$$

For example, for the case of the 24-bit Fibonacci 1-code (4.4) and 24-bit binary code (4.3) we have the following ranges of number's representations for the 24-bit Fibonacci 1-code (4.4) and the 24-bit binary code (4.3):

$$F_{25} = 75025,\ 2^{24} = 16777216. \tag{4.34}$$

By using (4.33) and (4.34), we can calculate the *potential error-detection ability* of the 24-bit Fibonacci 1-code (4.4) as follows:

$$S_d = 1 - \frac{75025}{16777216} = 0.99553\ (99.553\%). \tag{4.35}$$

### 4.7.3. *Fibonacci Parity Code*

In order to improve the potential error-detection ability of the Fibonacci 1-code (4.4), given by (4.33), we can use the so-called *Fibonacci Parity Code* (FPC) by adding the *parity bit* (PB) $a_{par}$ to the *minimal form* (MF) of the initial Fibonacci representation (*minimal form*):

$$\underbrace{a_n a_{n-1} \ldots a_i \ldots a_2 a_1}_{MF}\ \underbrace{a_{par}}_{PB}. \tag{4.36}$$

The *Fibonacci Parity Code* (FPC) significantly improves the error-detection ability of the Fibonacci 1-code from quantitative and qualitative points of view. In this case, the main qualitative feature of the

FPC is ensuring the 100% detection of all odd-bit errors, in particular, the single-bit errors. The addition of PB to the MF of the 24-bit Fibonacci 1-code (4.4) hasn't change the number of the *allowed* binary combinations in the FPC (4.36), which reminds equal to $F_{n+1}$, but the number of all bits of the FPC (4.36) is increasing on 1 bit. For this case the number of all possible code combinations is increasing to $2^{n+1}$. It follows from these arguments that the *potential error-detection ability* of the FPC (4.36) is calculated by the formula:

$$S_d\left(FPC\right)=1-\frac{F_{n+1}}{2^{n+1}}. \qquad (4.37)$$

For example, for the case of the initial 24-bit Fibonacci 1-code (4.4) we have the following ranges for the FPC (4.36) and the 25-bit binary code (4.3):

$$F_{25}=75025,\ 2^{25}=33554432. \qquad (4.38)$$

By using (4.37), we can calculate the *error-detection ability* of the 25-bit FPC (4.36) as follows:

$$S_d(FPC)=1-\frac{75025}{33554432}=0.998\left(99.8\%\right). \qquad (4.39)$$

This means that the 25-digit FPC (4.36) can provide the continuous detection of errors in Fibonacci micro-controller or microprocessor at various stages of storage, transmission and processing of data with the error detection coefficient (4.39).

### 4.7.4. *Fibonacci code for synchronization control*

A distinction of the bits 1 and 0 in sequential systems of data transmission leads to a serious technical problem, called *problem of synchronization*. To solve this problem, we need to use special synchronization signals, called *clock-signals*. One of the effective ways for improving the synchronization of transmitted data (without special clock-signals of synchronization) is to use *self-synchronization codes* (see diagram B in Fig. 4.4).

The diagrams A and B in Fig. 4.4 represent the widespread method for transmitting binary data. Switching two electrical currents gives the transmission of bits: the high level of the current $I_{high}$ and the low level of the current $I_{low}$. The switching from one level to another is performed only for the bits of 0; the switching is not performed for the bits of 1. Diagram A shows the electrical signal, which is generated during the transmission of the classic binary code combination without self-synchronization mechanism. The greatest difficulty in the transmission of the signal arises when the transmitted binary sequence contains a long "packet," consisting of the bits of 1.

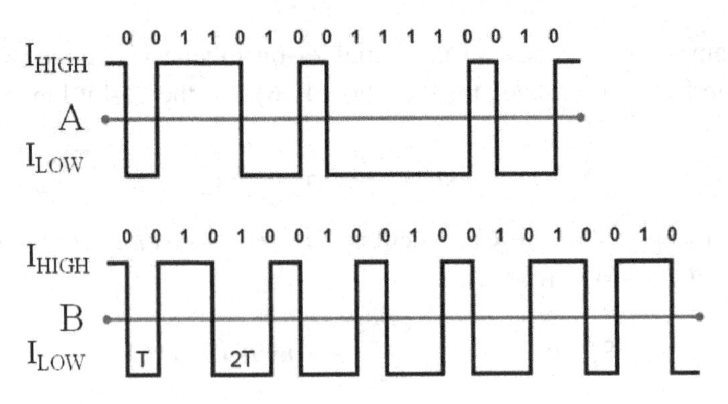


Figure 4.4. Fibonacci code as self-synchronization code.

The problem of distinction of the bits of 1 and 0 in this method is greatly simplified, if we impose certain restrictions on the length of the "packets," consisting of the consecutive bits of 1. The simplest solution is to use the codes, in which two bits of 1 do not appear together. This restriction is a basic property for the *minimal form* of the Fibonacci 1-code (5.4). For the first time, American engineer W.H. Kautz paid attention to such use of the Fibonacci code for synchronization control [82].

Diagram B shows the example of the formation of the digital signal by using the Fibonacci 1-code (4.4). Denote by $T$ the period of the original code sequence. Analysis of the digital signal, generated from the *minimal form* of the Fibonacci 1-code (4.4), shows that this digital signal contains only two pulse durations $T$ and $2T$. This digital signal is called *dual-frequency signal*. It is easy to form clock-signals from the dual-frequency signal. This fact is a confirmation of the self-synchronization property of the Fibonacci 1-code (4.4).

## 4.8. Concept of the Fibonacci arithmetical processor for noise-immune computations

### 4.8.1. *Noise-immune computations*

In modern computer science there is a need in the fault-tolerant and noise-immune processors. What is a distinction between these two important ways of designing the high-reliable computer and processors? It is well known that the computer program is realized through a processor. The processor consists of flip-flops, which are connected with combinative logic. In this case the realization of the program is reduced to the flip-flops switching. Unfortunately, it is impossible to eliminate computer errors, arising as the result of the fault functioning of computer elements. Nevertheless, it is necessary to distinguish two types of errors in computer elements. The first type is the so-called *constant failures*, when the elements "fall outside the order" constantly. The second type is the so-called *alternate failures*, when the elements "fall outside the order" temporarily, that is, in the accidental time moments, while in some other moments the computer elements are functioning correctly. The second type of failures is called *malfunction*. Processor's malfunctions appear under influence of the different internal and external noise factors in computer elements and their electrical circuits. Thus, the fault-tolerant computers are intended for the elimination of the *constant failures* that may appear in processor and its elements and units during their exploration. The noise-immune processors and computers are intended for the elimination of the "*malfunctions*" that may appear in computer elements during their exploration.

It is clear that the problem of designing the noise-immune computer and processor is actual problem of modern computer science. For example, many modern cryptosystems are based on the computations in very large finite fields. The hardware realization of such computational units or processors requires thousands of logic gates. It is very difficult and costly to develop these kinds of processors, which always yield error-free results. It means that a problem of the development of the processors for noise-immune computations is a very important problem of the reliable cryptosystem design.

It was proved experimentally that the intensity of the malfunctions (or random failures) in computer elements in the switching regime is bigger in two or three exponents of the intensity of the elements, which are in the stable states. It follows from this reasoning's, the *malfunctions of triggers in the switching regime is the most probable reason of the unreliable functioning of computer processors*. That is why, designing the self-checking automatic machines, which can provide the effective detection of triggers' malfunctions in the regime of their switching, is one of the most actual problems of the noise-immune computer and processor designing.

### 4.8.2. *Checking the basic micro-operations*

The basic idea of designing self-checking Fibonacci processor consists in the following. It is necessary to develop the effective system of checking the basic micro-operations in process of their fulfilment.

Let us demonstrate now a possibility of the realization of this idea by using the above *basic micro-operations* (*convolution*, *devolution*, *replacement* and *absorption*), used in the noise-immune Fibonacci arithmetic.

We pay attention to the following "technical" peculiarity of the above *basic micro-operations*. At the register interpretation of these micro-operations, each micro-operation may be realized by means of the inversion of the flip-flops, involved into the micro-operation. This means that each micro-operation is realized technically by means of flip-flops' switching.

Let us evaluate now a potential ability of the *basic micro-operations* to detect errors, which may appear during the micro-operations realization. As it is well-known, the potential error-detection ability is determined by the ratio between the number of the detectable errors and the general number of all possible errors. Let us explain the essence of our approach to the detection of errors in the above micro-operations on the example of the micro-operation of *convolution*:

$$011 \Rightarrow 100. \tag{4.40}$$

The *convolution* is fulfilled for the 3-digit binary code combination (4.40). It is clear that there are $2^3 = 8$ possible transitions, which can arise at the fulfillment of the micro-operation (4.40). Note that the only one of them, given by (4.40), is a *correct*, that is, *unmistakable* transition. The code combinations

$$\{011, 100\}, \tag{4.41}$$

which are involved in the *unmistakable* transition (4.40), are called *allowed* code combinations for the *convolution*. The other code combinations, which can appear during the *convolution* (4.40)

$$\{000, 001, 010, 101, 110, 111\}, \tag{4.42}$$

are *prohibited* code combinations.

The idea of the error detection consists in the following. If during the fulfillment of the micro-operation (4.40), one of the *prohibited* code combinations (4.42) appears, this is the indication of error. Note that if the erroneous transition

$$011 \Rightarrow 011, \tag{4.43}$$

when the *allowed* code combination 011 passes on into the same allowed code combination 011, we can interpret this transition as the case of *undetectable error*.

Let us consider now the different erroneous situations, which can appear at fulfillment of the micro-operation (4.40):

$$011 \Rightarrow \begin{Bmatrix} 0 & 1 & 1 \\ 0 & 0 & 0 \\ 0 & 0 & 1 \\ 0 & 1 & 0 \\ 1 & 0 & 1 \\ 1 & 1 & 0 \\ 1 & 1 & 1 \end{Bmatrix}. \tag{4.44}$$

Among them only the erroneous transition (4.43) is *undetectable*, because the code combination 011 is the *allowed* code combination. All the other *erroneous* transitions (4.44) are *detectable*.

Let us analyze the transition (4.43) from the arithmetical point of view. It is clear that the essence of the erroneous transition (4.43) consists in the *repetition* of the same code combination 011. If we analyze this transition from the arithmetical point of view, we can see that this transition does not destroy the numerical information and does not influence the outcome of the arithmetical operations. Hence, the *erroneous* transition (4.43) does not belong to the errors of *catastrophic character*. It can delay maybe only the data processing. All the other erroneous transitions from (4.44) are destroying the numerical information and hence can lead to the *errors of catastrophic character*.

The main conclusion consists in the fact that *the set of the "catastrophic" code combinations from* (4.42) *coincides with the set of the detectable code combinations* from (4.44). This means that all the "catastrophic" transitions for the *convolution* are *detectable*. We emphasize once again that the undetectable transition (4.43) does not destroy numerical information and, therefore, from the arithmetical point of view cannot belong to the erroneous transitions of catastrophic character. This *undetectable* transition is delaying only the data processing.

Thus, we can design by using this idea the computer device for the fulfillment of the *convolution* with the absolute (i.e. 100%) potential ability to detect all catastrophic transitions.

We can do the similar conclusion for other *basic micro-operations*. But the fulfillment of some data processing algorithm in the Fibonacci processor, based on the *basic micro-operations*, is reduced to the sequential fulfillment of the certain basic micro-operations on each computation step. **Because the checking of circuits for realization of the *basic micro-operations* has the "absolute" error-detecting ability, a possible design of the arithmetical self-checking Fibonacci processor is in the noise-immune Fibonacci processor at the flip-flop's switching.**

### 4.8.3. *The hardware realization of the noise-immune Fibonacci processor*

The noise-immune Fibonacci-processor is based on the principle of "cause-effect," described in article [23]. The essence of the principle consists in the following. The initial information (the "*cause*"), which is subject to the data processing, is transformed into the "*result*" by using some micro-operations. After that we transform the "*result*" (the "*effect*") to the initial information (the "*cause*") and then check that the "*effect*" fits to its "*cause.*" For example, at the fulfillment of the *convolution* for the binary combination 011 (the "*cause*"), we get the new binary combination 100 (the "*effect*"), which is necessary for the fulfillment of the *devolution.* Analogously the correct fulfillment of the *devolution* leads to the condition for the *convolution.* It follows from this consideration that the micro-operations of the *convolution* and *devolution* are mutually checked.

These conclusions are true for all the above *basic micro-operations*, represented in the table (4.28). For the "register interpretation" the correspondence between the "cause" and the "effect" is realized by using the "checking flip-flop". The "*cause*" sets up the "checking flip-flop" into the state of 1 and the correct fulfillment of the micro-operation (the "*effect*" fits to the "*cause*") overthrows the "checking flip-flop" into the state 0. If the "effect" does not fit to the "*cause*" (the micro-operation is fulfilled incorrectly), then the "checking flip-flop" remains in the state 1, which indicates the error.

If we analyze the "causes" and the "effects" for every basic micro-operation, we can determine that every "effect" is the inversion of its "cause," that is, all micro-operations could be realized by means of the inversion of the flip-flops, involved into the micro-operation.

The block diagram of the Fibonacci device for the realization of the principle of "cause-effect" is shown in Fig. 4.5. The device in Fig. 4.5 consists of the *information* and *check* registers, which are connected by means of the logic "*cause*" and "*effect*" circuits. The code information, entering the *information* register through the "Input," is analyzed by the logic "*cause*" circuit.

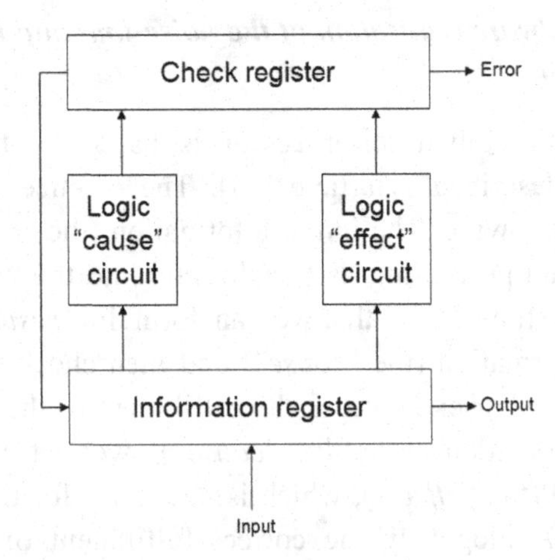


Figure 4.5. The block diagram of the Fibonacci device for the realization of the principle of the "cause-effect".

Suppose that we need to fulfill the *convolution* for the binary combination in the information register. Let some flip-flops $T_{k-1}$, $T_k$, $T_{k+1}$ of the *information* register be in the state 011, i.e. the condition for the *convolution* is satisfied for this group of flip-flops. Then, the logic "*cause*" circuit (the logic circuit for the *convolution* for this example) results in writing the logic 1 into the corresponding flip-flop $T_k$ of the *check* register. The written logic 1 is resulting into the inversion of the flip-flops $T_{k-1}$, $T_k$, $T_{k+1}$ of the *information* register by using the back connection, that is, their new states are 100. This means that the condition for the *devolution* is satisfied for this group of the flip-flops. Then, the logic "*effect*" circuit (the logic circuit for the *devolution* for this example) analyzes the states of the flip-flops $T_{k-1}$, $T_k$, $T_{k+1}$ of the *information* register and overthrows the same flip-flop $T_k$ of the *check* register to the initial state of 0. Overthrowing the flip-flop $T_k$ of the *check* register into the initial state of 0 confirms that the "*cause*" (011) fits to its "*effect*" (100), that is, the micro operation of the *convolution* is correct.

Hence, if we get the code word of 00…0 in the check register after the end of all micro-operations, this means that all "*causes*" fit to their "*effects*," that is, all the micro- operations are correct. If the *check* register contains at least one logic 1 in some flip-flop, this means that at least one basic micro-operation is not correct. The logic 1's in the flip-flops of the *check* register cause the error signal of 1 at the output "Error" of the device in Fig. 4.5. The signal of 1 at the output "Error" prohibits the use of the data on the "Output" of the Fibonacci device on Fig. 4.5.

The most important advantage of the check principle of the "cause-effect", which is realized in Fibonacci device in Fig. 4.5, is the detection of error in the moment of its appearance. The correction of error in the micro-operation is realized by the repetition of this micro-operation.

Hence, *the above approach, based on the principle of the "cause-effect," permits to detect and then to correct data by means of repetition of all "catastrophic" errors, arising in the moment of flip-flop switching with 100% guarantee.*

A more detailed description of all the benefits of this principle of implementating the noise-immune Fibonacci processor is given in article [23].

The article stresses that "*this approach can lead to designing a new class of high-reliable computers and processors, which provide a significant increase of the reliability of information processing in computer systems and the creation of new methods of information processing.*"

## 4.9. Boolean realization of the original Fibonacci element basis

### 4.9.1. *The register for reducing the Fibonacci code to the minimal form*

The *convolution* and *devolution* devices play an important role in implementing arithmetical operations in the Fibonacci 1-code. They can be designed on the base of the binary register, having special logic circuits to perform *convolutions* and *devolutions*. Each digit of the register contains a binary flip-flop and logic elements. The operations of *convolution* (011 = 100) and *devolution* (100 = 011) can be performed by means of the inversion of the flip-flops.

One of the possible variants of the *convolution* register or the *device for reducing the Fibonacci code to the minimal form* is shown in Fig. 4.6.

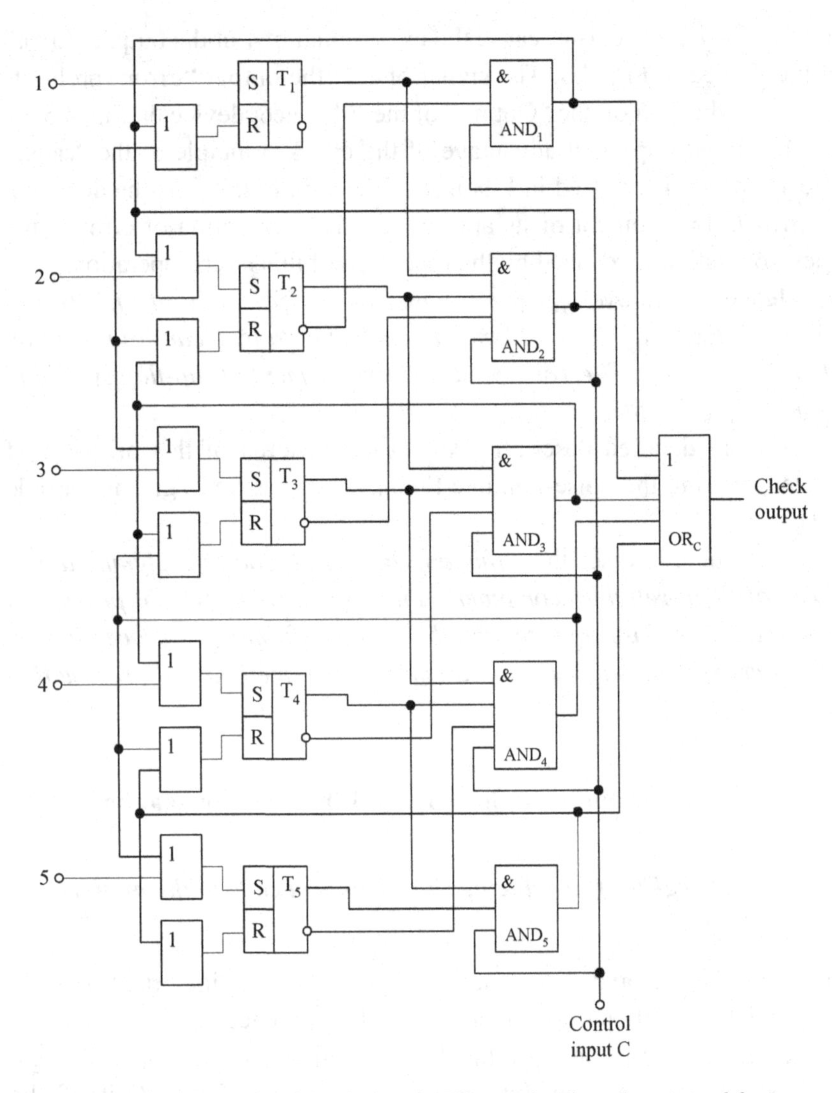


Figure 4.6. The device for reducing the Fibonacci code to the *minimal form*.

The device in Fig. 4.6 consists of the five *R-S*-triggers and the logic elements *AND*, *OR*, which are used to perform the *convolutions*. The *convolutions* are fulfilled by means of the logic elements $AND_1$–$AND_5$ and corresponding logic elements *OR*, standing before the *R*- and *S*-inputs of the flip-flops. The logic element $AND_1$ fulfills the *convolution* of the first digit to the second digit. Its two inputs are connected with the direct output of the flip-flop $T_1$ and the inverse output of the flip-flop $T_2$. The third input is connected with the synchronization input $C$. The logic element $AND_1$ analyzes the states $Q_1$ and $Q_2$ of the flip-flops $T_1$ and $T_2$. If $Q_1 = 1$ and $Q_2 = 0$, this means that the *convolution* condition is satisfied for the first and second digits. The synchronization signal $C = 1$ causes the appearance of the logic 1 at the output of the element $AND_1$. The latter causes switching the flip-flops $T_1$ and $T_2$. This leads to the *convolution* (01 = 10).

The logic element $AND_k$ of the $k$th digit ($k = 2, 3, 4, 5$) performs the *convolution* of the ($k-1$)th and $k$th digits to the ($k+1$)th digit. Its three inputs are connected with the direct outputs of the flip-flops $T_{k-1}$ and $T_k$ and the inverse output of the flip-flop $T_{k+1}$. The 4th input is connected with the synchronization input $C$. The logic element $AND_k$ analyzes the states $Q_{k-1}$, $Q_k$, and $Q_{k+1}$ of the flip-flops $T_{k-1}$, $T_k$, and $T_{k+1}$. If $Q_{k-1} = 1$, $Q_k = 1$, and $Q_{k+1} = 0$, this means that the *convolution* condition is satisfied. The synchronization signal $C = 1$ causes switching the flip-flops $T_{k-1}$, $T_k$, and $T_{k+1}$. The *convolution* of the corresponding digits (011 = 100) is over. Note that all elements $AND_1$–$AND_5$ are connected through the common element $OR_c$ with the check output of the *convolution* register.

The device for reducing the Fibonacci code to the *minimal form* in Fig. 6.6 operates in the following manner. The input code information is sent to the information inputs 1–5 of the *convolution* register and enters the $S$-inputs of the flip-flops through the corresponding logic elements $OR$. Let the initial state of the *convolution* register be as follows:

$$\begin{matrix} 5 & 4 & 3 & 2 & 1 \\ 0 & 1 & 0 & 1 & 1. \end{matrix}$$

It is clear that the condition for the convolution is satisfied for the first, second and third digits. The first synchronization signal $C$=1 is due to the transition of the convolution register to the new state:

$$\begin{matrix} 5 & 4 & 3 & 2 & 1 \\ 0 & 1 & 1 & 0 & 0. \end{matrix}$$

Here the condition of *convolution* is satisfied for the third, fourth and fifth digits. The next synchronization signal $C = 1$ is due to the transition of the *convolution* register to the following state:

5 4 3 2 1

1 0 0 0 0.

The *convolution* is over and the binary word is represented in the *minimal form*.

One may estimate the maximal delay time of the "convolution" operation for the $n$-digit *convolution* register. It is clear that we have to estimate the maximal time of the *convolution* delay for the following situation:

0 1 1 1 1 1

1 0 1 1 1 1

1 0 1 0 0 1

1 0 1 0 1 0.

We can see from this example that for even $n$ the maximal number of the sequential *convolutions* is equal to $\frac{n}{2}$.

The analysis of the logic circuit in Fig. 4.6 shows that the delay time of one *convolution* is defined as the sum of the *R-S*-flip-flop delay time $\tau_T$ and the delay time $\tau_e$ of the two sequential logic elements *AND*, *OR*, i.e.

$$\tau = \tau_T + 2\tau_e.$$

It follows from this consideration that the maximal *convolution* delay time is equal to:

$$\tau_C = \frac{n}{2}\left(\tau_T + 2\tau_e\right).$$

### 4.9.2. *The "convolution" register as self-checking device*

The outputs of the logic elements $AND_1$–$AND_5$ of the *convolution* registers in Fig. 5.6 are connected with the register check output through the common logic element *OR*. This output plays an important role as the check output of the *convolution* register.

It follows from the functioning principle of the *convolution* register that logic 1 appears on the check output only for two situations:

(1) The binary code word, written in the *convolution* register, is not *minimal form*. This means that the condition of *convolution* is satisfied at least for one triple of the adjacent flip-flops of the *convolution* register. This causes the appearance of logic 1 at the output of the corresponding element *AND*. Hence, in this case the appearance of logic 1 at the check output of the *convolution* register indicates the fact that the *convolution* process is not over. This means that there is a possibility to terminate the *convolution* process by means of observing the check output of the *convolution* register.

(2) The appearance of the constant logic 1 at the check output is an indication of the constant fault in the *convolution* register. Hence, the *convolution* register plays a role of self-checking device.

### 4.9.3. *The device for checking minimal form*

The *minimal form* is the main checking form for Fibonacci arithmetic. The combinative logic circuit for checking the *minimal form* is shown in Fig. 4.7. This logic circuit consists of $n$ logic elements *AND*. Their outputs are connected with the inputs of the common logic element *OR*. If the initial binary word has two adjacent binary 1's or the binary 1 in the lower digit, there appears the logic 1 at least on one output of the logic elements *AND*. It results in the appearance of logic 1 at the output of the common logic element *OR* and this logic 1 is the error indicator.

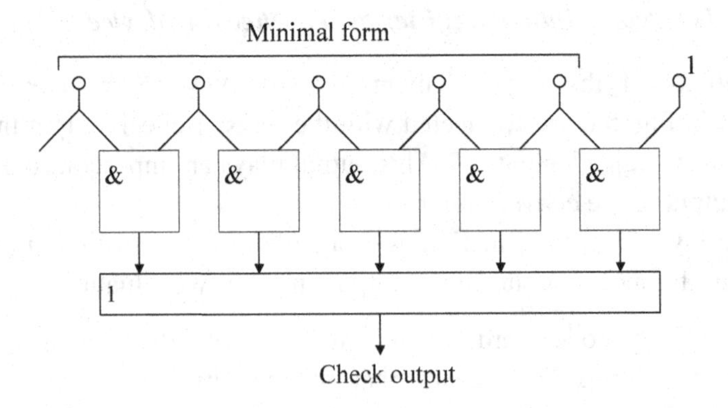


Figure 4.7. The logic circuit for checking the minimal form.

## 4.10. Fibonacci counter based on the *minimal form*

The researches in the field of new computer arithmetic's and their technical realizations are continuing intensively in the 21st century. First of all, it is appropriate to recall the two new scientific achievements that are of certain interest for computer science. In this connection, we should point on the *"golden" ternary mirror-symmetrical arithmetic* [28, 36], which is the original synthesis of *Bergman's system* and *Brusentsov's Ternary Principle.* The Fibonacci counter, based on the *minimal form* [83], is the next modern invention in this field. It should be noted that the great contribution of Professor Alexey Borisenko (Sumy University, Ukraine) in this invention.

### 4.10.1. *Mathematical justification*

Let us consider the formula (4.4) for the Fibonacci 1-code. First of all, we make one important remark. The Fibonacci counter, described below, is based only on the use of *minimal forms,* which are the *allowed* Fibonacci representations. Other Fibonacci representations are *forbidden* representations and their appearance is the indication of errors. As the least digit of the *minimal form* is identically equal to 0, in this case we can exclude the least digit of the Fibonacci 1-code (4.4) and consider the "truncated" Fibonacci 1-code with the weights of digits 1, 2, 5, 8, …, $F_n$, that is,

$$N = a_n F_n + a_{n-1} F_{n-1} + \ldots + a_i F_i + \ldots + a_3 F_3 + a_2 F_2. \qquad (4.45)$$

For example, the 5-bit "truncated" Fibonacci 1-code (4.45) looks as follows:

$$N = a_6 \times 8 + a_5 \times 5 + a_4 \times 3 + a_3 \times 2 + a_2 \times 1. \qquad (4.46)$$

Table 4.8 shows the sequential change of the *minimal forms* in the 5-digit "truncated" Fibonacci 1-code (4.46) and transitions between adjacent *minimal forms*, corresponding to the natural numbers $N$ and $N+1$.

Table 4.8. The *minimal forms* of the "truncated" Fibonacci 1-code.

| $i$ | 6 | 5 | 4 | 3 | 2 | $i$ | 6 | 5 | 4 | 3 | 2 |
|---|---|---|---|---|---|---|---|---|---|---|---|
| $N / F_i$ | 8 | 5 | 3 | 2 | 1 | $N / F_i$ | 8 | 5 | 4 | 3 | 2 |
| 0 | 0 | 0 | 0 | 0 | 0 | 7 | 0 | 1 | 0 | 1 | 0 |
| 1 | 0 | 0 | 0 | 0 | 1 | 8 | 1 | 0 | 0 | 0 | 0 |
| 2 | 0 | 0 | 0 | 1 | 0 | 9 | 1 | 0 | 0 | 0 | 1 |
| 3 | 0 | 0 | 1 | 0 | 0 | 10 | 1 | 0 | 0 | 1 | 0 |
| 4 | 0 | 0 | 1 | 0 | 1 | 11 | 1 | 0 | 1 | 0 | 0 |
| 5 | 0 | 1 | 0 | 0 | 0 | 12 | 1 | 0 | 1 | 0 | 1 |
| 6 | 0 | 1 | 0 | 0 | 1 | | | | | | |

Analysis of Table 4.8 allows identifying three typical transitions, arising at the change of the number $N$ on the number $N+1$:

1. Table 4.9 shows the **first typical transition**.

Table 4.9. The first typical transition at the change $N$ on $N+1$.

| $n$ | $n-1$ | … | $2k+3$ | $2k+2$ | $2k+1$ | $2k$ | $2k-1$ | $2k-2$ | $2k-3$ | … | 4 | 3 | 2 |
|---|---|---|---|---|---|---|---|---|---|---|---|---|---|
| $F_n$ | $F_{n-1}$ | … | $F_{2k+3}$ | $F_{2k+2}$ | $F_{2k+1}$ | $F_{2k}$ | $F_{2k-1}$ | $F_{2k-2}$ | $F_{2k-3}$ | … | $F_4$ | $F_3$ | $F_2$ |
| $a_n$ | $a_{n-1}$ | … | $a_{2k+3}$ | $a_{2k+2}$ | $a_{2k+1}$ | $a_{2k}$ | $a_{2k-1}$ | $a_{2k-2}$ | $a_{2k-3}$ | … | $a_4$ | **0** | **0** |
| $a_n$ | $a_{n-1}$ | … | $a_{2k+3}$ | $a_{2k+2}$ | $a_{2k+1}$ | $a_{2k}$ | $a_{2k-1}$ | $a_{2k-2}$ | $a_{2k-3}$ | … | $a_4$ | **0** | **1** |

The next 1 is written in the lowest digit of the Fibonacci 1-code (4.4) (that is, to the digit with the index 2) only if the bits of the first two digits (with the indices 2 and 3) are equal to 0 (see in bold).

The change of the Fibonacci representation of number $N$ in Table 4.9 on the Fibonacci representation, corresponding to the next number $N + 1$ (when the next bit 1 enters to the input of the Fibonacci counter) is realized according to the following **rule 1**: the bit of 1 is recorded into the digit with the index 2; herewith the new Fibonacci representation of the number $N + 1$ takes the minimal form, represented in the bottom line of Table 4.9. This rule corresponds to the transitions: 0→1, 3→4, 5→6, 8→9, 11→12 in Table 4.8.

2. Table 4.10 shows the **second typical transition**.

Table 4.10. The second typical transition at the change $N$ on $N + 1$.

| $n$ | $n-1$ | ... | $2k+3$ | $2k+2$ | $2k+1$ | $2k$ | $2k-1$ | $2k-2$ | $2k-3$ | ... | 4 | 3 | 2 |
|---|---|---|---|---|---|---|---|---|---|---|---|---|---|
| $F_n$ | $F_{n-1}$ | ... | $F_{2k+3}$ | $F_{2k+2}$ | $F_{2k+1}$ | $F_{2k}$ | $F_{2k-1}$ | $F_{2k-2}$ | $F_{2k-3}$ | ... | $F_4$ | $F_3$ | $F_2$ |
| $a_n$ | $a_{n-1}$ | ... | $a_{2k+3}$ | **0** | **0** | 1 | 0 | 1 | 0 | ... | 1 | 0 | 1 |
| $a_n$ | $a_{n-1}$ | ... | $a_{2k+3}$ | **0** | **1** | 0 | 0 | 0 | 0 | ... | 0 | 0 | 0 |

We can see that the Fibonacci representation of number $N$ in Table 4.10 has specific group of digits, starting from the digit with index 2 and ending by the digit with index $2k + 2$. This group of digits consists of two parts:

1) The first group starts from the digit with index 2 and ends with the digit with index $2k$. For this group the lower digit, that is, the digit with index 2 is equal to 1 and other digits are the alternating digits with bits 1 and 0, according to the property of *minimal form*; herewith the bits of 1 are in all the digits with even indices: 2, 4, 6, …, $2k - 2$, $2k$.
2) In the second group, the two digits with the indices $2k + 1$ and $2k + 2$ have bit 0 (see in bold). Note that the remaining group of digits with the indices from $2k + 3$ to $n$ satisfies the condition of the *minimal form* (two bits 1 together do not meet).

Then the change of the Fibonacci representation of number $N$ in Table 4.10 on the Fibonacci representation, corresponding to the next number $N + 1$ (when the next bit 1 enters the input of the Fibonacci counter), is realized according to the following **rule 2**: the bit of 1 is recorded into the digit with index $2k + 1$ and all the remaining digits with indices from 2 to $2k$ are *nulled*; herewith the new Fibonacci representation of number $N + 1$ takes the *minimal form*, represented in the bottom line of Table 4.10.

It is easy to prove that the Fibonacci representation in the bottom line of Table 4.10 corresponds to the natural number $N + 1$. This follows from property (1.25), which can be rewritten as follows:

$$F_2 + F_4 + F_6 + \ldots + F_{2k} = F_{2k+1} - 1. \tag{4.47}$$

3. Table 4.11 shows the **third typical transition**.

Table 4.11. The third typical transition at the change $N$ on $N + 1$.

| $n$ | $n-1$ | $\cdots$ | $2k+2$ | $2k+1$ | $2k$ | $2k-1$ | $2k-2$ | $2k-3$ | $2k-4$ | $\cdots$ | 4 | 3 | 2 |
|---|---|---|---|---|---|---|---|---|---|---|---|---|---|
| $F_n$ | $F_{n-1}$ | $\cdots$ | $F_{2k+2}$ | $F_{2k+1}$ | $F_{2k}$ | $F_{2k-1}$ | $F_{2k-2}$ | $F_{2k-3}$ | $F_{2k-4}$ | $\cdots$ | $F_4$ | $F_3$ | $F_2$ |
| $a_n$ | $a_{n-1}$ | $\cdots$ | $a_{2k+3}$ | **0** | **0** | 1 | 0 | 1 | 0 | $\cdots$ | 0 | 1 | 0 |
| $a_n$ | $a_{n-1}$ | $\cdots$ | $a_{2k+3}$ | **0** | **1** | 0 | 0 | 0 | 0 | $\cdots$ | 0 | 0 | 0 |

We can see that this Fibonacci representation of number $N$ in Table 4.11 has specific group of digits, starting from the digit with index 2 and ending by the digit with the index $(2k + 1)$. This group of digits consists of two parts:

1) For this first group, the digit with index 2 is equal to 0 and other digits are the alternating digits with bits 1 and 0; herewith the bits of 1 are in all the digits with the odd indices: $3, 5, 7, \ldots, 2k-3, 2k-1$.
2) The second group consists of the two digits with the indices $2k$ and $2k + 1$, which have bits 0 in the both digits (see in bold). Note that the remaining group of digits with the indices from $2k + 2$ to $n$ satisfies the condition of the *minimal form* (two bits of 1 together do not meet).

Then the change of the Fibonacci representation of number $N$ on the Fibonacci representation, corresponding to the next number $N + 1$ (when the next bit 1 enters the input of the Fibonacci counter), is realized according to the following **rule 3**: the bit of 1 is recorded in the digit with the index $2k$ and all the remaining digits with the indices from 2 to $2k$-1 are *nulled*; herewith the new Fibonacci representation of the number $N+1$ takes the form, represented in the last row of Table 4.11.

Let's prove that the Fibonacci representation in the last row of Table 4.11 corresponds to the positive integer $N+1$. It is clear that the weight of the digit with index $2k$ is equal to $F_{2k}$. On the other hand, according to (1.24) the sum of the Fibonacci numbers with odd indices 1, 3, 5, …, $2k-1$ is equal to $F_{2k}$, that is,

$$F_1 + F_3 + F_5 + \ldots + F_{2k-1} = F_{2k}. \quad (4.48)$$

Let's recall that we consider the "truncated" Fibonacci 1-code (4.45), where the digit with index 1 is absent. Taking into consideration this fact, we can rewrite the identity (4.48) as follows:

$$F_3 + F_5 + \ldots + F_{2k-1} = F_{2k} - F_1 = F_{2k} - 1. \quad (4.49)$$

Thus, the numerical equivalent of the *nulled* digits in the above Fibonacci representation is equal to $F_{2k} - 1$, whereas the numerical equivalent of the digit with index $2k$ is equal $F_{2k}$. It follows from this consideration that the new Fibonacci representation (see the last row of Table 4.11), in fact, corresponds to the natural number $N+1$.

### 4.10.2. *Structural realization of the Fibonacci counter*

A new algorithm for the Fibonacci counter, based on the *minimal form*, shows that the transition from the Fibonacci representation of the number $N$ into the Fibonacci representation of the next number $N+1$ is carried out in one clock cycle, which includes the record of 1 into the appropriate digit, and 0 in other digits. This creates prerequisites for the creation of high-speed Fibonacci counter. But most importantly, this counter uses only the *minimal forms* for the representation of numbers; these forms are the *allowed* Fibonacci representations and the main "checking" forms of the Fibonacci code.

This means that the Fibonacci counter, based on *minimal form*, allows solving at once two important tasks:
- improving the speed of counter;

- providing continuous checking the counter in all stages of data transformation in the counter.

For ease of description of the structural circuit of the Fibonacci counter, we renumber the digits in the "truncated" Fibonacci code (4.45) and use the Fibonacci code, in which the numeration of digits begins with the first digit:

$$N = a_n F_{n+1} + a_{n-1} F_n + \ldots + a_{i-1} F_i + \ldots + a_2 F_3 + a_1 F_2. \quad (4.50)$$

Analysis of the above Fibonacci counting algorithm shows that we should have 5 operational blocks for the realization of the Fibonacci counter: *Register*, *Disposition Block*, *Block for Analysis*, *Block for Error Checking*, and *Block for Zero Installation*. These blocks allow designing a high-speed and noise-immune Fibonacci counter, based on the *minimal form*. Its noise-immunity is achieved due to the presence of *forbidden* states of the counter, but the high speed is achieved as a result of the absence of carry-overs that are needed in the binary counters, and also the micro-operations of *convolutions* and *devolutions* that are used in the known Fibonacci counters [84]. These properties give some benefits compared with classical binary counters and the known Fibonacci counters [84].

Let us consider the operation of the Fibonacci counter, based on the *minimal form* on the example of the 5-bit Fibonacci counter (see Fig. 4.8).

The Fibonacci counter in Fig. 4.8 consists of the *Register*, formed by 5 two-state RS-Flip-Flops FF1-FF5 with the control logic circuits AND1–AND5, *Block for Analysis*, consisting of the logic gates AND6–AND10, *Disposition Block*, consisting of the logic gates AND11–AND15, *Block for Error Checking*, consisting of the logic gates AND16–AND20 and OR6, *Block for Zero Installation*, consisting of the logic gates AND21 and OR1-OR5. All these blocks have a regular structure and the number of its digits can be increased easily. To increase the capacity of the Fibonacci counter, for example, on one digit (the sixth digit), one just needs to connect the corresponding outputs of the sixth digits to the inputs 1–9. By analogy, we can design the seventh, eighth digits and so on up to the *n*th digit of the Fibonacci counter.

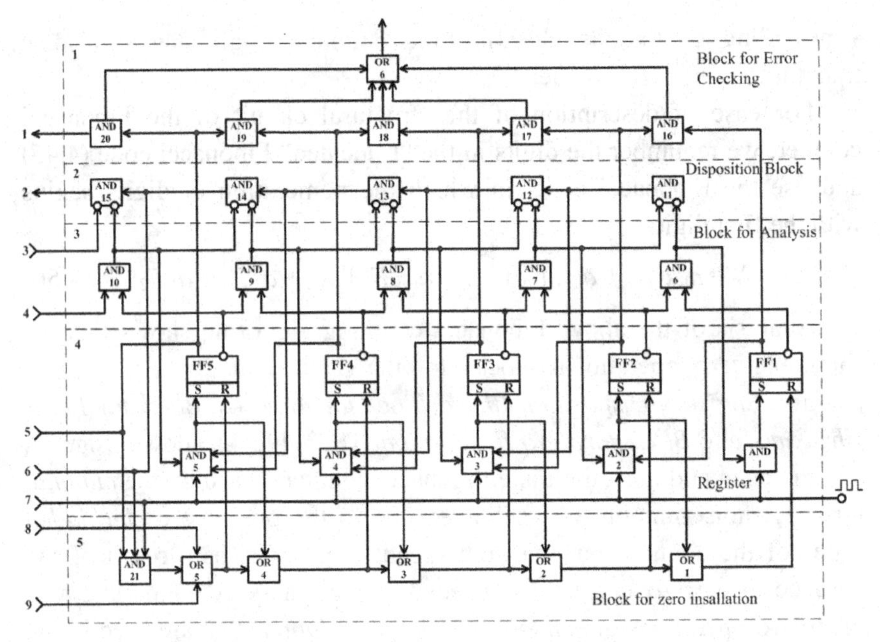


Figure 4.8. The Fibonacci counter, based on the *minimal form*.

The *Register* with the logic gates AND1-AND5 is needed to memorize the counter states. In the example on Fig. 4.8 the counter uses the RS-Flip-Flops FF1-FF5, as the most simple, although we may use any other kinds of Flip-Flops, for example, JK-Flip-Flops. The outputs of the RS-Flip-Flops FF1-FF5 are connected to the inputs of the logic gates AND6-AND10 of the *Block for Analysis*, but the inputs of RS-Flip-Flops FF1-FF5 are connected with the *Block for Zero Installation* and the bus clock 7 for syncronization signals *C*.

The *Block for Analysis* checks the outputs of the pairs of adjacent Flip-Flops of the *Register*, starting from FF1 and FF2, which can be in three *allowed* states 00, 01 or 10. Let us recall that the state 11 for the two adjacent flip-flops is *forbidden*, in accordance with the main property of the *minimal form*. In order to identify the *forbidden* states, we introduced into the Fibonacci counter the *Block for Error Checking*, which indicates the error state of the Fibonacci counter. To analyze the location of the 1's in the Fibonacci counter, we introduced the *Disposition Block*. The *Disposition Block* together with the *Block for Analysis* checks that the Flip-Flops of the *Register* go to the next correct state of the counter. The *Block for Zero Installation* sets all the Flip-Flops of the *Register* into state 0 for the two cases. The first case, when any Flip-Flop of the *Register* is switched over into the state 1; herewith, all the Flip-Flops on the right should be switched over into the state 0. For the second case, the *Block for Zero Installation* sets all the Flip-Flops of the *Register* into state 0 after the end of the counting cycle.

### 4.10.3. *Operation of the Fibonacci counter*

***The Initial state***

The Fibonacci counter in Fig. 4.8 operates as follows. First, the Fibonacci counter is in the initial state, when all the Flip-Flops FF1-FF5 of the *Register* are installed in the state 0. For this case, the logic signals 0 from the direct outputs of the Flip-Flops FF1-FF4, according to Fig. 4.8, enter to the inputs of the logic gates AND2-AND5, respectively. Note that the inverse outputs of the Flip-Flops FF1-FF2 are connected with the inputs of the logic gate AND6, and the inverse outputs of the Flip-Flops FF2-FF3 are connected with the inputs of the logic gate AND7 and so on up to the logic gate AND10, as is shown in Fig. 4.8; this means that on the outputs of the logic gates AND6–AND10 the signals 1 are formed. The logic signals of 1 from the outputs of the logic gates AND6–AND10 enter to the corresponding inputs of the logic gates AND1–AND5, respectively.

***The First step***

According to Fig. 4.8, the first clock pulse enters to the inputs of the logic gates AND1–AND5. However, if the Fibonacci counter is in the state 0 = 00000, only the Flip-Flop FF1 through the logic gate AND1 can be switched over into state 1 because all the other logic gates AND2–AND5 are blocked by the 0-signals, entered from the direct outputs of the Flip-Flops FF2-FF5. As a result, the Fibonacci counter is installed into the new state 00001, corresponding to the positive integer 1.

***The Second step***

The transition of the Flip-Flop FF1 into state 1 leads to the following situation. The signal 0 appears on the output of the logic gate AND6 and then on the input of the logic gate AND1. This signal is forbidding the transition of the next clock pulse to the S-input of the Flip-Flop FF1. The logic gate AND2, on the contrary, is open for the transition of the next clock pulse to the S-input of the Flip-Flop FF2. This means that the next (second) clock pulse is switching over the Flip-Flop FF2 into the state 1, while the second clock pulse, through the logic gate OR1, is switching over the Flip-Flop FF1 to the state 0. As a result, we get a new state of the Fibonacci counter 00010, corresponding to the natural number 2.

***The Third step***

Let us consider the situation with the logic gate AND3 after the second step. This logic gate has 4 inputs. As the Fibonacci counter is in state 00010, this means that the signal 1 from the direct output of the Flip-Flop FF2 is entering the first input of the logic gate AND3. Because the Flip-Flops FF3 and FF4 are in the state 0, this means that the signal 1 from the output of the logic gate AND8 is entering the second input of the logic gate AND3. Finally, let us consider the situation with the logic gates AND6, AND7 and AND11. Because the Fibonacci counter is in state 00010, this means that the input signals of the logic gates AND6, AND7 are equal to 0 and the input signal of the logic gate AND11 is equal to 1. Thus, the signal 1 is entering the third input of the logic gate AND3. This means that the third clock pulse, through the logic gate AND3, enters the S-input of the Flip-Flop FF3 and is switching over it into the state 1. Besides, this logic signal 1 enters, through the logic gates OR2 and then OR1, to the R-inputs of the Flip-Flops FF2 and FF1 and switching over them into state 0. As a result, we get the new state of the Fibonacci counter 00100, corresponding to the natural number 3.

***The Fourth step***

When the Fibonacci counter is in state 00100, the logic signals of 1 appear on the inverse outputs of the Flip-Flops FF2 and FF1. These logic signals lead to the appearance of signal 1 on the output of the logic gate AND6. On the other hand, signal 1 from the output of the logic gate AND6 enters the input of the logic gate AND11 of the *Disposition Block*. This leads to the appearance of logic 0 at the output of the logic gate AND11. The signal 0 from the output of the logic gate AND11 leads to the appearance of signals 0 on the outputs of the logic gates AND12, AND13, AND14. These signals 0 enter the inputs of the logic gates AND3, AND4, AND5. Finally, the signal 0 on the inverse output of the Flip-Flop FF3, which is in state 1, causes the logic signal of 0 on the output of the logic gate AND7 and on the input of the logic gate AND2. This means that the next (fourth) clock pulse does not change the states of the Flip-Flops FF2-FF5. However, the Flip-Flop FF1 is switched over into state 1 by the fourth clock signal. As a result, the counter is established in state 00101 what corresponds to the natural number 4.

### ***The Fifth step***

Let us consider now, how is fulfilled the transition from the number 4 = 00101 to the next number 5 = 01000. For the situation 00101 the Flip-Flops FF1 and FF3 are in state 1. This means that the signals of 0 appear on the outputs of the logic gates AND6–AND8. These signals enter the inputs of the logic gates AND1–AND3. Due to the signals of 0 on the outputs of the logic gates AND7, AND8, signal 1 appears on the output of the logic gate AND12, and then this signal enters the input of the logic gate AND4. In addition, signal 1 enters another input of the logic gate AND4 from the direct output of the Flip-Flop FF3, which is in state 1. Also, signal 1 enters the third input of the logic gate AND4 from the output of the logic gate AND9. As a result, the next (fifth) clock pulse sets Flip-Flop FF4 in state 1, and then, through logic gates OR3-OR1, sets the Flip-Flops FF3, FF2, FF1 into state 0. As a result, the Fibonacci counter goes into the state 01000, corresponding to the number 5 = 01000.

### ***The Sixth step***

As the Flip-Flops FF1, FF2, FF3 are in the states of 0, the signals 0 enter the corresponding inputs of the logic gates AND2–AND4 from the direct outputs of these Flip-Flops. Besides, signals 1 enter the inputs of the logic gates AND6, AND7 from the inverse outputs of the Flip-Flops FF1, FF2, FF3, which are in states of 0. The signals of 1 enter the inverse inputs of the logic gates AND11 and contribute to the appearance of signal 0 on the output of the logic gate AND11. This signal 0 contributes to the appearance of signal 0 on the output of the logic gate AND12 and then AND13. The signal 0 enters the input of the logic gate AND5 from the output of the logic gate AND13. Because the Flip-Flops FF1, FF2 are in states of 1, this causes the appearance of signal 1 on the output of the logic gate AND6. This signal 1 enters the input of the logic gate AND1. Thus, only the Flip-Flop FF1 can be switched over by the clock pulse. The next (sixth) clock signal is switching over the Flip-Flop FF1 into state 1 and does not change the states of the Flip-Flops FF2-FF5. As a result, the Fibonacci counter goes to the state of 01001, corresponding to the number 6 = 01001.

***The Seventh step***

Further, by analogy to the next (seventh) clock pulse by transferring the Fibonacci counter into the state 01010, corresponding to the number 7 = 01010.

***The Eighth step***

For this situation, signals 1 appear on the inputs of the logic gate AND5 and the next (eight) clock pulse sets the Fibonacci counter into state 10000, corresponding to number 8 = 10000.

***The Last steps***

This process will continue up to the transition of the Fibonacci counter into state 10101, corresponding to the maximal number 12 = 10101, which can be represented with the 5-bit Fibonacci code in the *minimal form*. For this case, two signals of 1 enter the inputs of the logic gate AND21 from the direct input of the Flip-Flop FF5 and the logic gate AND14 of the *Disposition Block*. The next clock pulse, through the logic gates AND21 and OR5-OR1, sets the Flip-Flops FF1 - FF5 into the initial state 00000. After that, the counter is ready for the new counting cycle.

***Increasing the Fibonacci counter capacity***

As it is mentioned above, the additional inputs and outputs 1-9 are intended for increasing the Fibonacci counter capacity on the left and on the right. For that, we can use the standard set of the logic gates for one digit of the counter. This standard structure consists of the one two-input logic gate OR of the *Block for Zero Installation*, the one Flip-Flop and the one logic gate AND of the *Register*, the one two-input logic gate AND of the *Block for Analysis*, the one two-input logic gate AND of the *Disposition Block*, and the one two-digit logic gate AND of the *Block for Error Checking*.

**Block for Error Checking**

The *Block for Error Checking* is, in fact, an external block for the counter and does not affect the counting algorithm. Its task is to indicate errors in the counter. The random appearance of the two adjacent 1's in the Flip-Flops of the *Register* leads to the fact that the *error signal* 1 appears on one of the logic gates AND of the *Block for Error Checking*. This *error signal* passes through the logic gate OR6 to the output of the *Block for Error Checking*.

### 4.10.4. *The advantages of Fibonacci counter, based on the minimal form*

The main advantage of the given Fibonacci counter compared with the well-known Fibonacci counters [84] is that the counter is operating only in the *minimal form*. All other forms of number representation in the Fibonacci code are *forbidden*, and their occurrence is the indication of errors, which may appear in the Fibonacci counter under various internal and external influences.

Fibonacci counting device has a fairly high speed. This is achieved due to the fact that the carry-overs from the digit to digit are absent. This improves its speed in comparison with the known binary error-correcting counters and the known Fibonacci counters [84], based on the *convolutions* and *devolutions*.

As it follows from Fig. 4.8, the proposed counter has a high homogeneity of logical elements and connections between them which is important for the microelectronics and nano-electronics technology. This is another advantage of the Fibonacci counter.

Thus, the proposed counter is noise-immune, sufficiently fast, and has a regular structure and high informational reliability. Therefore, it makes sense to recommend this counter for designing digital structures for mission-critical applications.

## 4.11. USA researches in Fibonacci computer field

It is necessary to note that along with the Soviet studies on "Fibonacci Arithmetic" and "Fibonacci computers", in the same period, similar studies have been fulfilled in the United States (University of Maryland) under the scientific supervision of Prof. Robert Newcomb [85–89].

Figure 4.13. Professor Robert Newcomb, University of Maryland (in the center) with colleagues.

The studies of the American, Soviet, and Ukrainian scientists in this field are confirmation of the fact that, since 1970s, the notions of "Fibonacci code," "Fibonacci arithmetic" and "Fibonacci computer" have become widely known in the world scientific and technical literature.

# Chapter 5

# Codes of the Golden $p$-Proportions and Their Applications in Computer Science and "Golden" Metrology

## 5.1. Definition of the codes of the golden $p$-proportions

The classic binary system has the following interpretation. Consider the infinite set of the binary standard line segments:

$$\{2^n, 2^{n-1}, \ldots, 2^i, \ldots, 2^0, 2^{-1}, \ldots, 2^{-i}, \ldots, 2^{-(n-1)}, 2^{-n}\} \tag{5.1}$$

where $i = 0, \pm1, \pm2, \pm3, \ldots$ .

By using (5.1), we can represent every real number $A$ as follows:

$$A = \sum_i a_i 2^i, \tag{5.2}$$

where $a_i$ is the binary numeral of the $i$th digit; $2^i$ is the weight of the $i$th digit; $i = 0, \pm1, \pm2, \pm3, \ldots$ .

The digit weights are connected by the two binary "arithmetical" properties:

$$2^i = 2 \times 2^{i-1} \text{ (binary multiplicative property)}, \tag{5.3}$$

$$2^i = 2^{i-1} + 2^{i-1} \text{ (binary additive property)}, \tag{5.4}$$

which underlie the "binary arithmetic".

The *binary multiplicative property* (5.3) is the basis of the binary multiplication, left- and right-shifts of the binary words, and the representation of numbers with "floating point." The *binary additive property* (5.4) underlies the binary summation and binary subtraction (through the *inverse* and *additional* codes).

The binary code of the real number $A$, which is determined by (5.1), assumes the following generalization. Let us consider the set of the following standard line segments:

$$\{\Phi_p^n, \Phi_p^{n-1}, \ldots, \Phi_p^0 = 1, \Phi_p^{-1}, \ldots, \Phi_p^{-k}, \ldots\}, \tag{5.5}$$

where $\Phi_p$ is the golden $p$-ratio, the positive root of the golden $p$-ratio equation (1.66) $x^{p+1} = x^p + 1$.

By using (5.5), we can get the following positional method of the real number representation:

$$A = \sum_i a_i \Phi_p^i, \tag{5.6}$$

where $A$ is positive real number, $a_i \in \{0,1\}$ is the binary numeral of the $i$th digit; $\Phi_p^i$ is the weight of the $i$th digit; $\Phi_p$ is the base of the numeral system (5.6), $i = 0, \pm 1, \pm 2, \pm 3, \ldots, p = 0, 1, 2, 3, \ldots$ is the given integer.

## 5.2. Partial cases of the codes of the golden $p$-proportions

Let us consider now more in detail the above method of the real number representation, which is given by (5.6). First of all, we can note that (5.6) sets forth a theoretically infinite number of the binary positional representations of real numbers, because every $p$ = 0, 1, 2, 3, … "generates" its own method of the binary positional number representation in the form (5.6).

The base of numeral system is one of the fundamental notions of positional numeral system. The analysis of the sum (5.6) shows that the base of the numeral system (5.6) is the golden $p$-ratio $\Phi_p$, the positive root of the golden $p$-ratio equation (1.66) $x^{p+1} = x^p + 1$. That is why, the representation of the real numbers $A$ in the form (5.6) is called the *codes of the golden p-proportions of real number A* [15].

Note that except for the case $p = 0$ ($\Phi_{p=0} = 2$) all the other golden $p$-proportions $\Phi_p$ are irrational numbers. It follows from this fact that the *codes of the golden p-proportions* (5.6) are the binary numeral systems with irrational bases $\Phi_p$ for the cases $p > 0$.

Note that for $p = 0$, the *codes of the golden p-proportions* (5.6) are reduced to the classic binary code (5.2) and for $p = 1$ to *Bergman's system* (3.1). It is clear that Bergman's system (3.1) has the most practical significance because this numeral system with irrational base $\Phi = \frac{1+\sqrt{5}}{2}$ is the simplest for technical realization.

## 5.3. Conversion of numbers from traditional numeral systems to Bergman's system

### 5.3.1. *A table method*

There are two general methods of number conversion from one numeral system to other. The first method is called a *table method* and the second method is called an *analytic method.* The *table method* is based on the preliminary construction in computer of a special table for the *codes of the golden p-proportions*. This method can be realized in computer by means of a special constant electronic memory. In this case the *codes of the golden p-proportions* of the number $N$ are kept in the memory by the address, which is the classic binary representation of the number $N$. For the case $p = 1$ (*Bergman's system*) such table has the following form (Table 5.1).

Note that the top row of Table 5.1 contains the golden ratio powers of the kind $\Phi^k$ ($k = 0, \pm1, \pm2, \pm3, \ldots$).

Table 5.1. Table method of number conversion into Bergman's system.

| Address $N$ | $\Phi^4$ | $\Phi^3$ | $\Phi^2$ | $\Phi^1$ | $\Phi^0$ | $\Phi^{-1}$ | $\Phi^{-2}$ | $\Phi^{-3}$ | $\Phi^{-4}$ |
|---|---|---|---|---|---|---|---|---|---|
| $0 = 0000$ | 0 | 0 | 0 | 0 | 0. | 0 | 0 | 0 | 0 |
| $1 = 0001$ | 0 | 0 | 0 | 0 | 1. | 0 | 0 | 0 | 0 |
| $2 = 0010$ | 0 | 0 | 0 | 1 | 0. | 0 | 1 | 0 | 0 |
| $3 = 0011$ | 0 | 0 | 1 | 0 | 0. | 0 | 1 | 0 | 0 |
| $4 = 0100$ | 0 | 0 | 1 | 0 | 1. | 0 | 1 | 0 | 0 |
| $5 = 0101$ | 0 | 1 | 0 | 0 | 0. | 1 | 0 | 0 | 1 |
| $6 = 0110$ | 0 | 1 | 0 | 1 | 0. | 0 | 0 | 0 | 1 |
| $7 = 0111$ | 1 | 0 | 0 | 0 | 0. | 0 | 0 | 0 | 1 |
| $8 = 1000$ | 1 | 0 | 0 | 0 | 1. | 0 | 0 | 0 | 1 |
| $9 = 1001$ | 1 | 0 | 0 | 1 | 0. | 0 | 1 | 0 | 1 |
| $10 = 1010$ | 1 | 0 | 1 | 0 | 0. | 0 | 1 | 0 | 1 |
| $11 = 1011$ | 1 | 0 | 1 | 0 | 1. | 0 | 1 | 0 | 1 |

### 5.3.2. *Conversion of fractional numbers*

Consider now the analytic method of the conversion of fractional numbers to their "golden" representations in Bergman's system (3.1). This method is widely used in the classic numeral systems. Its essence consists in the fulfillment of some arithmetical operations in the initial numeral system for obtaining numerals in the new numeral system.

Suppose that the "golden" representation of the fractional number $A$ has the following form:

$$A = a_{-1}\Phi^{-1} + a_{-2}\Phi^{-2} + \ldots + a_{-n}\Phi^{-n} = 0.a_{-1}a_{-2}\ldots a_{-n}. \quad (5.7)$$

Suppose that the fractional number $A$ is represented in the *minimal form*. Then, by multiplying the fractional number (5.7) by base $\Phi$, we get the following result:

$$A \times \Phi = a_{-1} + a_{-2}\Phi^{-1} + \ldots + a_{-n}\Phi^{-n+1} = a_{-1}.a_{-2}\ldots a_{-n}, \quad (5.8)$$

where $a_{-1}$ is the integer part of the product $A \times \Phi$ and the sum

$$A_1 = a_{-2}\Phi^{-1} + \ldots + a_{-n}\Phi^{-n+1} = 0.a_{-2}\ldots a_{-n} \quad (5.9)$$

is the fractional part of the product (5.8).

Thus, it follows from this consideration that after the first multiplication of the initial fractional number (5.7) on base $\Phi$, the integer part of the product (5.8) is the binary numeral $a_{-1}$ of the "golden" representation of the fractional number (5.7).

By multiplying the fractional number (5.9) on base $\Phi$, we get the following result:

$$A_1 \times \Phi = a_{-2} + a_{-3}\Phi^{-1} + \ldots + a_{-n}\Phi^{-n+2} = a_{-2}.a_{-3}\ldots a_{-n}. \quad (5.10)$$

The analysis of (5.10) shows that the second multiplication on base $\Phi$ leads us to the binary numeral $a_{-2}$ of the "golden" representation of the initial fractional number (5.7).

By continuing the multiplication process $n$ times, we get the "golden" representation of the fractional number $A$.

**Example 5.1.** Convert the decimal fraction $\frac{1}{2}$ into its "golden" representation in Bergman's system.

**Solution:**

*First multiplication:*

$$\frac{1}{2}\times\Phi=\frac{1}{2}\times\frac{1+\sqrt{5}}{2}=\frac{1+\sqrt{5}}{4}\approx 0.809. \qquad (5.11)$$

As the integer part of the fractional number (5.7) is equal to 0, it follows from this that the first "golden" binary numeral of the decimal fraction $\frac{1}{2}$ is $a_{-1} = 0$.

*Second multiplication:*

$$\left(\frac{1}{2}\times\Phi\right)\times\Phi=\frac{1}{2}\times\Phi^2=\frac{1}{2}\times\frac{3+\sqrt{5}}{2}=\frac{3+\sqrt{5}}{4}\approx 1.309. \qquad (5.12)$$

As the integer part of the obtained product (5.12) is equal to 1, this means that $a_{-2} = 1$. It follows from this result that before the third multiplication it is necessary to subtract the number 1 from the number (5.12):

$$\frac{3+\sqrt{5}}{4}-1=\frac{\sqrt{5}-1}{4}=\frac{1}{2}\times\Phi^{-1}\approx 0.309.$$

*Third multiplication:*

$$\left(\frac{1}{2}\times\Phi^{-1}\right)\times\Phi^{1}=\frac{1}{2}=0.5.$$

As a result of the third multiplication, we got the fractional number 0.5. This means that the "golden" bit $a_{-3}$ of the decimal fraction $\frac{1}{2}$ is $a_{-3}=0$.

After the third multiplication we came to the initial fractional number $\frac{1}{2}$. It follows from this fact that further multiplication will result in the repetition of the obtained binary numerals $a_{-1} = 0$, $a_{-2} = 1$, $a_{-3} = 0$. Hence, the "golden" representation of the decimal fraction $\frac{1}{2}$ has the form of the following periodic fraction:

$$\frac{1}{2} = 0.010010010\ldots.$$

### 5.3.3. *Conversion of integers*

The analytic method to obtain the "golden" representation of integers is reduced to the sequential comparison of the initial number $N$ and arising here reminders with the powers of the golden ratio.

**Example 5.2.** Convert the integer number 4 into the "golden" representation in Bergman's system.
**Solution:**

By using the formula (1.43) $\Phi^n = \frac{L_n + F_n\sqrt{5}}{2}$, we can get the decimal equivalents (D.E.) for the powers of the golden ratio (see Table 5.2).

Table 5.2. Decimal equivalents for the powers of the golden ratio.

| $n$ | 3 | 2 | 1 | 0 | −1 | −2 | −3 |
|---|---|---|---|---|---|---|---|
| $\Phi^n$ | $\frac{4+2\sqrt{5}}{2}$ | $\frac{3+\sqrt{5}}{2}$ | $\frac{1+\sqrt{5}}{2}$ | 1 | $\frac{-1+\sqrt{5}}{2}$ | $\frac{3-\sqrt{5}}{2}$ | $\frac{-4+2\sqrt{5}}{2}$ |
| *D.E.* | 4.236 | 2.618 | 1.618 | 1 | 0.618 | 0.382 | 0.236 |

*The first step of conversion*

By comparing integer 4 with the D.E. of the golden ratio powers (Table 5.2), we find the pair of the powers $\Phi^3 = \frac{4+2\sqrt{5}}{2} = 4.236$ and $\Phi^2 = \frac{3+\sqrt{5}}{2} = 2.618$, which are connected with number 4 by the following non-equalities:

$$\frac{3+\sqrt{5}}{2} = 2.618 \le 4 < \frac{4+2\sqrt{5}}{2} = 4.236. \tag{5.13}$$

It follows from (5.13) that the binary numeral of the second digit of the "golden" representation of number 4 is $a_2 = 1$.

*The second step of conversion*

Represent 4 as follows:

$$4=\frac{3+\sqrt{5}}{2}+r_1. \tag{5.14}$$

Let us calculate the value of the remainder $r_1$ as follows:

$$r_1=4-\frac{3+\sqrt{5}}{2}=\frac{5-\sqrt{5}}{2}=1.382. \tag{5.15}$$

By comparing the reminder (5.15) to the golden ratio powers (Table 5.2), we found the next pair of the golden ratio powers, $\Phi^1=\frac{1+\sqrt{5}}{2}=1.618$ and $\Phi^0=1$, which are connected with the reminder $r_1 = 1.382$ by the following inequality:

$$1\le 1.382<\frac{1+\sqrt{5}}{2}=1.618. \tag{5.16}$$

It follows from (5.16) that the binary numeral of the 0th digit of the "golden" representation of number 4 is $a_0 = 1$.

*The third step of conversion*

Represent the first reminder $r_1$ as follows:

$$r_1=\frac{5-\sqrt{5}}{2}=1+r_2, \tag{5.17}$$

where the second reminder $r_2$ is equal:

$$r_2=r_1-1=\frac{5-\sqrt{5}}{2}-1=\frac{3-\sqrt{5}}{2}=0.382. \tag{5.18}$$

By comparing the second reminder (5.18) to the golden ratio powers (Table 5.2), we can find the next pair of the golden ratio powers, $\Phi^{-1}=\frac{-1+\sqrt{5}}{2}=0.618$ and $\Phi^{-2}=\frac{3-\sqrt{5}}{2}=0.382$, which are connected with the second reminder $r_2 = 0.382$ by the following inequality:

$$\frac{3-\sqrt{5}}{2}=0.382\leq 0.382<\frac{-1+\sqrt{5}}{2}=0.618. \quad (5.19)$$

It follows from (5.19) that the binary numeral $a_{-2}$ of the "golden" representation of number 4 is $a_{-2} = 1$.

*The fourth step of conversion*

Represent the second reminder $r_2=\frac{3-\sqrt{5}}{2}=0.382$ as follows:

$$r_2=\frac{3-\sqrt{5}}{2}=\frac{3-\sqrt{5}}{2}+r_3, \quad (5.20)$$

where the third reminder $r_3$ is equal:

$$r_3=r_2-\frac{3-\sqrt{5}}{2}=0. \quad (5.21)$$

As the third reminder $r_3 = 0$, this means that the conversion process is over and the conversion result is the following:

$$4 = 101.01. \quad (5.22)$$

Note that the above Examples 5.1 and 5.2 are the basis for the development of computer algorithm of number conversion to the "golden" representation.

It is important to emphasize that the algorithms of the number conversion is similar to the algorithms of the number conversion to the classic binary system.

## 5.4. The "golden" arithmetic

### 5.4.1. *The "golden" multiplicative and additive properties*

There are the following important properties, which connect the digit weights of the codes (5.6):

$$\Phi_p^i=\Phi_p\times\Phi_p^{i-1}\text{ (“golden” multiplicative property)} \quad (5.23)$$

$$\Phi_p^i = \Phi_p^{i-1} + \Phi_p^{i-p-1} \text{ (“golden” additive property)}, \tag{5.24}$$

where $p = 0, 1, 2, 3, \ldots$ and $i = 0, \pm 1, \pm 2, \pm 3, \ldots$ .

The properties (5.23), (5.24) are the basis of the “golden” arithmetic in the codes of the *golden p-proportions* (5.6).

From the point of view of the simplicity of the technical implementation, *Bergman's system* is the most interesting; this system is a special case ($p = 1$) of the codes of the golden $p$-proportions (5.6). For the case $p = 1$ the “golden” properties (5.23), (5.24) take the following forms:

$$\Phi^i = \Phi \times \Phi^{i-1} \text{ (“golden” multiplicative property, } p = 1), \tag{5.25}$$

$$\Phi^i = \Phi^{i-1} + \Phi^{i-2} \text{ (“golden” additive property, } p = 1), \tag{5.26}$$

where $i = 0, \pm 1, \pm 2, \pm 3, \ldots$ and $\Phi = \frac{1+\sqrt{5}}{2}$ (the golden ratio).

The “golden” identities (5.25), (5.26) underlie the “golden” arithmetic for *Bergman's system* $A = \sum_i a_i \Phi^i$.

### 5.4.2. *The “golden” summation and subtraction*

Let us compare the “*golden*” *additive properties* (5.24), (5.26) to the recurrent relation (1.57) for the Fibonacci $p$-numbers $F_p(n) = F_p(n-1) + F_p(n-p-1)$ and the recurrent relation (1.2) $F_n = F_{n-1} + F_{n-2}$ for the classic Fibonacci numbers, respectively (see Table 5.3).

Table 5.3. Comparison of the “golden” additive properties to the recurrent relations for Fibonacci $p$-numbers and classic Fibonacci numbers.

| $p$ | "Golden" additive properties | Recurrent relations for Fibonacci $p$ - numbers and classic Fibonacci numbers |
|---|---|---|
| $p \geq 0$ | $\Phi_p^i = \Phi_p^{i-1} + \Phi_p^{i-p-1}$ | $F_p(i) = F_p(i-1) + F_p(i-p-1)$ |
| $p = 1$ | $\Phi^i = \Phi^{i-1} + \Phi^{i-2}$ | $F_n = F_{n-1} + F_{n-2}$ |

Table 5.3 shows that the mathematical structures of the “*golden*” *additive property* (5.24) and the recurrent relation (1.57) for the

Fibonacci $p$-numbers as well the "*golden*" *additive property* (5.26) and the recurrent relation (1.2) for the classic Fibonacci numbers are similar.

As shown above, the recurrent relations for the Fibonacci $p$-numbers (1.57) and the classic Fibonacci numbers (1.2) are the basis of the original Fibonacci arithmetic, in particular, *Fibonacci summation* and *subtraction*, based on the four *basic micro-operations*. It follows from Table 5.3, that the same *basic micro-operations* can be used to fulfill the arithmetic operations of the "golden" summation and subtraction.

Let us consider the following examples of the "golden" summation and subtraction of the natural numbers, represented in Chapter 3 as follows:

$$\Phi\text{-}code\ (3.9)\ N = \sum_i a_i \Phi^i,$$

$$F\text{-}code\ (3.36)\ N = \sum_i a_i F_{i+1},$$

$$L\text{-}code\ (3.40)\ N = \sum_i a_i L_{i+1}.$$

Recall that all the codes (3.9), (3.36), (3.40) are the distinct "golden" codes of one and the same natural number $N$ in *Bergman's system* (3.9), and the bits $a_i \in \{0,1\}\ (i = 0, \pm1, \pm2, \pm3, \ldots)$ for all the codes (3.9), (3.36), (3.40) are coincident.

Recall also that these codes (3.9), (3.36), (3.40) have a very effective method of error detection, based on the so-called $Z$- and $D$-properties, based on Theorems 3.2 and 3.3.

For $i = 0, \pm1, \pm2, \pm3, \ldots$ the $Z$- and $D$-properties are the following:

***Z-property***

$$\text{For any } N = \sum_i a_i \Phi^i \text{ after substitution } F_i \to \Phi^i \text{ we have: } \sum_i a_i F_i \equiv 0 \quad (5.27)$$

***D-property***

$$\text{For any } N = \sum_i a_i \Phi^i \text{ after substitution } L_i \to \Phi^i \text{ we have: } \sum_i a_i L_i \equiv 2N \quad (5.28)$$

Let us consider now the following examples of the "golden" summation and subtraction of natural numbers, based on the *basic micro-operations* (4.28).

**Example 5.3.** Summarize the natural numbers 5 + 4, represented in the Φ-*code* (3.9).
**Solution:**

(1) Let us represent the natural numbers $N_1 = 5$ and $N_2 = 4$ in the Φ-*code* (3.9) by using the *minimal form* (see rows 5 and 6 in Table 5.4).

Table 5.4. "Golden" representations of the numbers $N_1 = 5$ and $N_2 = 4$ in the Φ-*code*.

| $i$ | → | 3 | 2 | 1 | 0. | −1 | −2 | −3 | −4 |
|---|---|---|---|---|---|---|---|---|---|
| $\Phi^i$ | → | $\Phi^3$ | $\Phi^2$ | $\Phi^1$ | $\Phi^0$ | $\Phi^{-1}$ | $\Phi^{-2}$ | $\Phi^{-3}$ | $\Phi^{-4}$ |
| $F_i$ | → | 2 | 1 | 1 | 0. | 1 | −1 | 2 | −3 |
| $\downarrow N / F_{i+1}$ | → | 3 | 2 | 1 | 1. | 0 | 1 | −1 | 2 |
| $N_1 = 5$ | = | 1 | 0 | 0 | 0. | 1 | 0 | 0 | 1 |
| $N_2 = 4$ | = | 0 | 1 | 0 | 1. | 0 | 1 | 0 | 0 |

<u>$N_1 = 5$</u>

According to the *F-code* interpretation (see rows 4 and 5 of Table 5.4, the number $N_1 = 5$ is equal to the following sum:

$$N_1 = 1\times3 + 0\times2 + 0\times1 + 0\times1 + 1\times0 + 1\times1 + 0\times(-1) + 1\times2 = 3 + 2 = 5.$$

By checking the "golden" representation of number $N_1 = 5$ by the *minimal form* (two bits of 1 together don't meet), we see that the "golden" representation of number $N_1 = 5$ is correct.

By checking the "golden" representation of number $N_1 = 5$ by the *Z-property* (see the row 3), we came to the following result:

$$1\times2 + 0\times1 + 0\times1 + 0\times0 + 1\times1 + 0\times(-1) + 0\times2 + 1\times(-3) = 2 + 1 - 3 = 0.$$

This means that the "golden" representation of number $N_1 = 5$ is correct.

$\underline{N_2 = 4}$

According to the *F-code* interpretation (see rows 4 and 5), the number $N_2 = 4$ is equal to the following sum:

$$N_2 = 0\times3 + 1\times2 + 0\times1 + 1\times1 + 0\times0 + 1\times1 + 0\times(-1) + 0\times(-2) = 2 + 1 + 1 = 4.$$

By checking the "golden" representation of number $N_2 = 4$ by the *minimal form,* we came to the conclusion that the "golden" representation of number $N_2 = 4$ is correct (two bits of 1 together don't meet).

By checking the "golden" representation of number $N_2 = 4$ by the *Z-property,* we came to the following result:

$$N_2 = 0\times2 + 1\times1 + 0\times1 + 1\times0 + 0\times0 + 0\times1 + 1\times(-1) + 0\times2 + 0\times(-3) = 1 - 1 = 0.$$

This means that the "golden" representation of number $N_2 = 4$ is correct.

(2) Let us fulfill the micro-operation of *replacement* (Table 5.5).

Table 5.5. The micro-operations of *replacement.*

| $i$ | → | 3 | 2 | 1 | 0 | 1 | −1 | 2 | −3 |
|---|---|---|---|---|---|---|---|---|---|
| $\downarrow N/\Phi^i$ | → | $\Phi^3$ | $\Phi^2$ | $\Phi^1$ | $\Phi^0$ | $\Phi^{-1}$ | $\Phi^{-2}$ | $\Phi^{-3}$ | $\Phi^{-4}$ |
| $N_1 = 5$ | = | 1 | 0 | 0 | 0. | 1 | 0 | 0 | 1 |
| + | | ↓ | | | | ↓ | | | ↓ |
| $N_2 = 4$ | = | 0 | 1 | 0 | 1 | 0 | 1 | 0 | 0 |
| $S = 5 + 4$ | = | 1 | 1 | 0 | 1. | 1 | 1 | 0 | 1 |

(3) Let us reduce the summation result $S = 5 + 4$ to the *minimal form*:

$$S = 5 + 4 = 1101.1101 = 10010.0101. \tag{5.29}$$

(4) *F-code* interpretation of the *minimal form* of the summation result (5.29) and its checking by the *minimal form* and by *Z-property* (Table 4.17).

Table 5.6. F-code interpretation and checking the summation result.

| $i$ | $\rightarrow$ | 4 | 3 | 2 | 1 | 0 | −1 | −2 | −3 | −4 |
|---|---|---|---|---|---|---|---|---|---|---|
| $\Phi^i$ | $\rightarrow$ | $\Phi^4$ | $\Phi^3$ | $\Phi^2$ | $\Phi^1$ | $\Phi^0$ | $\Phi^{-1}$ | $\Phi^{-2}$ | $\Phi^{-3}$ | $\Phi^{-4}$ |
| $F_i$ | $\rightarrow$ | 3 | 2 | 1 | 1 | 0 | 1 | −1 | 2 | −3 |
| $\downarrow N/F_{i+1}$ | $\rightarrow$ | 5 | 3 | 2 | 1 | 1 | 0 | 1 | −1 | 2 |
| $S=5+4$ | = | 1 | 0 | 0 | 1 | 0. | 0 | 1 | 0 | 1 |

According to the *F-code* interpretation (see rows 4 and 5 of Table 5.6), the sum $S = 5 + 4$ is the following:

$$S = 1\times5 + 0\times3 + 0\times2 + 1\times1 + 0\times1 + 0\times0 + 1\times1 + 0\times(-1) + 1\times2$$
$$= 5 + 1 + 1 + 2 = 9.$$

By checking the "golden" representation of the sum $S = 5 + 4$ by the *minimal form*, we conclude that the "golden" representation of the number $S = 5 + 4$ is correct (two bits of 1 together don't meet).

By checking the "golden" representation of the sum $S$ by the *Z-property* (see rows 3 and 5 of Table 5.6), we came to the following result:

$$1\times3 + 0\times2 + 0\times1 + 1\times1 + 0\times1 + 0\times0 + 0\times1 + 1\times(-1) + 0\times2 + 1\times(-3)$$
$$= 3 + 1 - 1 - 2 = 0.$$

This means that the "golden" representation of the sum $S = 5 + 4$ is correct.

**Example 5.3.** Subtract number 11 from number 3 in *Bergman's system*.
**Solution:**
(1) Let us represent the two numbers 3 and 11 in the *minimal form*:

$$3 = 00100.0100 \text{ and } 11 = 10101.0101 \quad (5.30)$$

and check them by the *Z-property* (Table 5.7).

Table 5.7. Representation of numbers 3 and 11 in the *minimal form.*

| $i$ | → | 4 | 3 | 2 | 1 | 0 | −1 | −2 | −3 | −4 |
|---|---|---|---|---|---|---|---|---|---|---|
| $\Phi^i$ | → | $\Phi^4$ | $\Phi^3$ | $\Phi^2$ | $\Phi^1$ | $\Phi^0$ | $\Phi^{-1}$ | $\Phi^{-2}$ | $\Phi^{-3}$ | $\Phi^{-4}$ |
| $F_i$ | → | 3 | 2 | 1 | 1 | 0 | 1 | −1 | 2 | −3 |
| $\downarrow N / F_{i+1}$ | → | 5 | 3 | 2 | 1 | 1 | 0 | 1 | −1 | 2 |
| $N_1 = 3$ | = | 0 | 0 | 1 | 0 | 0. | 0 | 1 | 0 | 0 |
| $N_1 = 11$ | = | 1 | 0 | 1 | 0 | 1. | 0 | 1 | 0 | 1 |

<u>***N₁ = 3***</u>

According to the *F-code* interpretation (see rows 4 and 5), $N_1 = 3$ is equal to the following sum:

$$N_1 = 0\times5 + 0\times3 + 1\times2 + 0\times1 + 0\times1 + 0\times0 + 1\times1 + 0\times(-1) + 0\times2$$
$$= 2 + 1 = 3.$$

By checking the "golden" representation of $N_1 = 3$ by the *Z-property*, we came to the following result:

$$N_2 = 0\times2 + 1\times1 + 0\times1 + 1\times0 + 0\times0 + 0\times1 + 1\times(-1) + 0\times2 + 0\times(-3)$$
$$= 1 - 1 = 0.$$

This means that the "golden" representation of $N_1 = 3$ is correct.

<u>***N₂ = 11***</u>

According to the *F-code* interpretation, $N_2 = 11$ is equal to the following sum:

$$N_2 = 1\times5 + 0\times3 + 1\times2 + 0\times1 + 1\times1 + 0\times0 + 1\times1 + 0\times(-1) + 1\times2$$
$$= 5 + 2 + 1 + 1 + 2 = 11.$$

By checking the "golden" representation of $N_2 = 11$ by the *Z-property*, we came to the following result:

$$N_2 = 1\times3 + 0\times2 + 1\times1 + 1\times0 + 0\times1 + 1\times0 + 0\times1 + 1\times(-1) + 0\times2$$
$$+1\times(-3) = 3 + 1 - 1 - 3 = 0.$$

This means that the "golden" representation of $N_2 = 11$ is correct.

(2) Let us fulfill the micro-operations of *absorption* (Table 5.8).

Table 5.8. The micro-operations of *absorption*.

| $i$ | $\rightarrow$ | 4 | 3 | 2 | 1 | 0 | −1 | −2 | −3 | −4 |
|---|---|---|---|---|---|---|---|---|---|---|
| $\downarrow N/\Phi^i$ | $\rightarrow$ | $\Phi^4$ | $\Phi^3$ | $\Phi^2$ | $\Phi^1$ | $\Phi^0$ | $\Phi^{-1}$ | $\Phi^{-2}$ | $\Phi^{-3}$ | $\Phi^{-4}$ |
| $N_1=3$ | = | 0 | 0 | 1 | 0 | 0. | 0 | 1 | 0 | 0 |
| – | $\rightarrow$ | | | $\updownarrow$ | | | | $\updownarrow$ | | |
| $N_2=11$ | = | 1 | 0 | 1 | 0 | 1. | 0 | 1 | 0 | 1 |
| $D=3-11$ | = | 1 | 0 | 0 | 0 | 1. | 0 | 0 | 0 | 1 |

Because the subtraction result $D = 3 - 11$ is in the bottom register, this means that the result $D$ has the "–" sign.

### 5.4.3. *Representation of numbers with the floating point*

Compare now the *codes of the golden p-proportions* (5.6) to the *Fibonacci p-codes* (4.1). These codes are similar by their mathematical structures. However, there is the following distinction between them:

(1) First of all, it is necessary to note that the *Fibonacci p-codes* (5.1) are intended for the representation of integers and demand less digits for representation one and the same range of number representation in comparing the *codes of the golden p-proportions* (5.6). For example, the decimal number 10 needs for its representation in *Bergman's system* the 9-digit binary code combination:

$$10 = 10100.0101.$$

For the Fibonacci 1-code we need only 6 binary digits for the representation of the same number 10:

$$10 = 100100.$$

Thus, we can give a preference to the Fibonacci $p$-codes for the code representation of integers.

(2) The next distinction between the *Fibonacci p-codes* (4.2) and the *codes of the golden p-proportions* (5.6) is connected with digit

weights. The digit weights of the *codes of the golden p-proportions* (5.6) are forming the geometric progression (5.5). This fact gives a possibility to fulfill left- and right-shifts of the "golden" representations. This corresponds to multiplication and division of the initial number by base $\Phi_p$ (the golden $p$-ratio). The "shift-property" of the golden $p$-ratio codes gives a possibility to represent numbers with the *floating point*. In fact, the decimal number 10 can be represented in *Bergman's system* with floating point as follows:

$$10_{10} = 10100.0101_{\Phi} = 0.101000101 \times \Phi^5. \tag{5.31}$$

The "golden" representation with floating point (5.31) consists of two parts. The first part is the *"golden" mantissa* of the decimal number 10

$$m(10) = 0.101000101 \tag{5.32}$$

and the second part is the golden ratio power $\Phi^5$. The number 5 is called the *"golden" exponent* of number 10.

### 5.4.4. *The "golden" multiplication*

The "golden" multiplication is based on the following trivial identity for the golden $p$-ratio powers:

$$\Phi_p^n \times \Phi_p^m = \Phi_p^{n+m}. \tag{5.33}$$

Table 5.9 follows from (5.33) for the *"golden" multiplication*, which is true for all codes of the golden $p$-proportions (5.6).

Table 5.9. Table of the "golden" multiplication.

| | | |
|---|---|---|
| $0 \times 0$ | = | 0 |
| $0 \times 1$ | = | 0 |
| $1 \times 0$ | = | 0 |
| $1 \times 1$ | = | 1 |

We can see that Table 5.9 for the "golden" multiplication coincides with the multiplication table, used in the classic binary arithmetic. This means that the "golden" multiplication is reduced to the classic binary multiplication, i.e. to the following rules:

(1) To form the partial products in accordance with Table 5.9.
(2) To summarize the partial products in accordance with the rule of the "golden" summation.

**Example 5.4.** Multiply the "golden" fractions $A = 0.010010$ and $B = 0.001010$ in *Bergman's system*.

**Solution:**

(1) Let us represent the "golden" fractional numbers $A = 0.010010$ and $B = 0.001010$ in the form with floating point:

$$A = 010010 \times \Phi^{-6},$$
$$B = 001010 \times \Phi^{-6}.$$

This means that the "golden" mantissas and "golden" exponents of numbers $A$ and $B$ are equal to the following, respectively:

$$m(A) = 010010;\ e(A) = -6 \text{ and } m(B) = 001010;\ e(B) = -6.$$

(2) Let us multiply the "golden" mantissas:

$$\begin{array}{rccccccccccc}
A= & & & & & & 0 & 1 & 0 & 0 & 1 & 0 \\
B= & & & & & & 0 & 0 & 1 & 0 & 1 & 0 \\
\hline
 & & & & & & 0 & 0 & 0 & 0 & 0 & 0 \\
 & & & & & 0 & 1 & 0 & 0 & 1 & 0 & \\
 & & & & 0 & 0 & 0 & 0 & 0 & 0 & & \\
 & & & 0 & 1 & 0 & 0 & 1 & 0 & & & \\
 & & 0 & 0 & 0 & 0 & 0 & 0 & & & & \\
 & 0 & 0 & 0 & 0 & 0 & 0 & & & & & \\
\hline
 & 0 & 0 & 0 & 1 & 0 & 1 & 1 & 0 & 1 & 0 & 0
\end{array}$$

(3) Let us reduce the product $A \times B$ to the *minimal form*:

$$A \times B = 00010110100 = 00100000100.$$

(4) Let us summarize the exponents:

$$e(A) + e(B) = (-6) + (-6) = -12.$$

Then we can represent the product $A \times B$ in the form with the floating point:

$$A \times B = 00100000100 \times \Phi^{-12}.$$

### 5.4.5. *The "golden" division*

For the creation of the rules of the "golden" division we can use an analogy between the classic binary arithmetic and the "golden" arithmetic. As it is well known, the classic binary division is reduced to the shift of the divisor and comparison of the shifted divisor with the divisible or with the intermediate divisible. If the divisible or the intermediate divisible is more than the shifted divisor, then the binary numeral 1 is written to the corresponding digit of the quotient, in the opposite case is written the binary 0. For the former case the shifted divisor is subtracted from the divisible or the intermediate divisible. One may show that these operations underlay the "golden" division too. Because the comparison of numbers is fulfilled over the numbers, represented in the *minimal form*, it follows from this consideration that a peculiarity of the "golden" division consists in the fact that the divisible and all the intermediate divisibles should be reduced to the *minimal form* on every step of the division.

Let us demonstrate now the "golden" division on the following example.

**Example 5.5.** Divide the "golden" number 5 = 1000.1001 (the divisible) by the "golden" number 10=10100.0100 (the divisor) in *Bergman's system*.

**Solution:**

Let us represent the numbers 5 and 10 in the form with the floating point:

$$m(A) = 5 \times \Phi^4 = 10001001; e(A) = 4, \tag{5.34}$$

$$m(B) = 10 \times \Phi^4 = 101000101; e(B) = 4. \tag{5.35}$$

Because their "golden" exponents are equal, that is, $e(A) = e(B)$, we can write:

$$5:10 = m(A): m(B),$$

that is, the division of the initial numbers 5:10 is reduced to the division of their "golden" mantissas $m(A)$: $m(B)$.

Let us divide now the mantissa $A$ by the mantissa $B$:

(1) If we compare the mantissa $m(A)$ to the mantissa $m(B)$, we find that $m(A) < m(B)$. This means that the result of the division is a proper fraction, that is, the binary numeral of the 0th digit of the quotient $Q$ is $a_0 = 0$.

(2) If we shift the "golden" mantissa $m(A)$ by 1 digit to the left, we get the following:

$$A_1 = 100010010. \tag{5.36}$$

(3) If we compare the number (4.86) to the mantissa $m(B)$, we find that $A_1 < m(B)$. This means that the binary numeral of the next digit of the quotient $Q$ is equal to $a_{-1} = 0$.

(4) By shifting the number (4.86) by 1 digit to the left, we get the following number:

$$A_2 = 1000100100. \tag{5.37}$$

(5) If we compare number (5.37) to the "golden" mantissa $m(B)$, we find that $A_2 \geq m(B)$. This means that the next binary numeral of the quotient $Q$ is $a_{-2} = 1$.

(6) Because we obtained the bit $a_{-2} = 1$, we need to subtract the "golden" mantissa $m(B)$ from the number $A_2$. As a result, we find the next intermediate result

$$A_3 = A_2 - m(B) = 1000100.1. \tag{5.38}$$

A peculiarity of (5.38) is that it has the bit 1 in its fractional part.

(7) By shifting the "golden" representation (5.38) on 1 digit to the left, we get a new intermediate result:

$$A_4 = 10001001. \tag{5.39}$$

Note that the golden representation (5.39) is equal to the "golden" mantissa $m(A)$. This means that starting from (5.39), the division process is repeated. Hence, the next binary numerals of the quotient will be equal: $a_{-3} = 0$, $a_{-4} = 0$, $A_{-5} = 1\ldots$ .

It follows from this consideration that the quotient has the following "golden" representation:

$$Q = 0.010010010...,$$

that is, for this case the quotient $Q = (5:10) = 0.5$ is represented in *Bergman's system* as a periodic fraction.

If we compare the "golden" arithmetic with the Fibonacci arithmetic and the classic binary arithmetic, we can find two peculiar properties of the "golden" arithmetic:

(1) The rules of the "golden" summation and subtraction coincide with the corresponding rules of the Fibonacci arithmetic.
(2) Similar to the classic binary code, the codes of the golden $p$-proportions (5.6) have the important arithmetical property to represent the numbers in the form with floating point.
(3) The rules of the "golden" multiplication and division coincide with the similar rules for the classic binary arithmetic.

Thus, the "golden" arithmetic is a peculiar synthesis of the Fibonacci and classic binary arithmetic.

## 5.5. Application of the codes of the golden $p$-proportions in digital metrology

### 5.5.1. *What is metrology?*

We can find the answer to this question in Russian and English Wikipedia's. The Russian Wikipedia gives the following definition of *metrology* [90]:

*"Metrology is the science of measurement, methods and tools to ensure their unity and ways to achieve the required accuracy. The subject of metrology is to extract quantitative information about the properties of objects with the specified accuracy and reliability; the regulatory framework for this — metrological standards."*

The English Wikipedia gives the following definition of *metrology* [91]:

*"Metrology is defined by the International Bureau of Weights and Measures (BIPM) as the science of measurement, embracing both*

*experimental and theoretical determinations at any level of uncertainty in any field of science and technology.”*

In this section we describe the *golden resistive divisors* as the basis of self-correcting analog-to-digit and digit-to-analog converters.

For the first time, this original theory was described in author’s 1978 article *Digital Metrology on the base of the Fibonacci codes and Golden Proportion Codes* [16].

### 5.5.2. *The “binary” resistive divisor*

In measurement practice the so-called *resistive divisors*, intended for the division of electric currents and voltages in the given ratio, are widely used. One of the variants of such divisor is shown in Fig. 5.1.

The resistive divisor in Fig. 5.1 consists of the “horizontal” resistors of the kind *R*1 and *R*3 and the “vertical” resistors *R*2. The resistors of the divisor are connected between themselves by the “connecting points” 0, 1, 2, 3, 4. Each point is connected to three resistors, which are forming together the resistive section. Note that Fig. 5.1 shows the resistive divisor, which consists of 5 resistive sections. In general, a number of resistive sections can be extended *ad infinitum*.

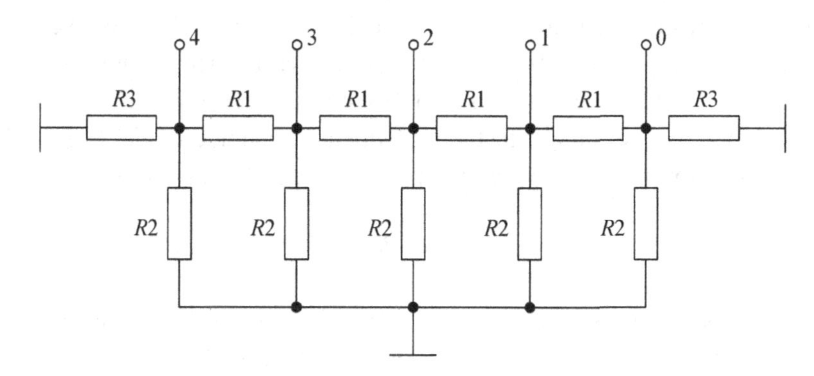


Figure 5.1. The resistive divisor.

First of all, we note that the parallel connection of the resistors *R*2 and *R*3 to the right of the “connecting point” 0 and to the left of the “connecting point” 4 can be replaced by the equivalent resistor with the resistance, which can be calculated according to the well-known electric law on parallel connection of two resistors $R2$ and $R3$ (see Fig. 5.1):

$$R_{e1} = \frac{R2 \times R3}{R2 + R3}. \tag{5.40}$$

Then, it is easy to calculate the equivalent resistance of the resistive section to the right of the "connecting point" 1 and to the left of the "connecting point" 3:

$$R_{e2} = R1 + R_{e1}. \tag{5.41}$$

In dependence on the choice of the resistance values of the resistors $R1$, $R2$, $R3$, we can get the different coefficients for the current or voltage division. Let us consider now the "binary" resistive divisor,. For this case the resistive divisor consists of the following resistors: $R1 = R$, $R2 = R3 = 2R$, where $R$ is some standard resistance value. For this case the expressions (5.40), (5.41) take the following values:

$$R_{e1} = R;\ R_{e2} = 2R. \tag{5.42}$$

Then, taking into consideration (5.42), we can prove that the equivalent resistance of the "binary" resistive divider to the left or to the right of any "connecting point" 0, 1, 2, 3, 4 is equal to $2R$. This means that the equivalent resistance of the resistive divisor in the "connecting points" 0, 1, 2, 3, 4 can be calculated as the resistance of the parallel connection of the three resistors of the value $2R$. By using the electric circuit laws, we can calculate the equivalent resistance of the "binary" resistive divisor in each "connecting point" 0, 1, 2, 3, 4 as follows:

$$R_{e3} = \frac{2}{3}R. \tag{5.43}$$

Let us connect now the generator of the standard electric current $I$ to one of the "connecting points", for example, to point 2. Then according to Ohm's law the following electric voltage appears in this point:

$$U = \frac{2}{3}RI. \tag{5.44}$$

Let us calculate now the electric voltages in the "connecting points" 3 and 1, which are adjacent to point 2. It is easy to show that the voltage transmission coefficient between the adjacent "connecting points" is equal to $\frac{1}{2}$. This means that the "binary" resistive divisor fits very well to the binary system and this fact is a cause of wide use of the "binary" resistive divisor in Fig. 5.1 in modern "binary" digit-to-analog and analog-to-digit converters.

### 5.5.3. *The "golden" resistive divisors and their electric properties*

Let us take the values of the resistors of the resistive divisors in Fig. 5.1 as follows:

$$R1 = \Phi_p^{-p} R;\ R2 = \Phi_p^{p+1} R;\ R3 = \Phi_p R, \tag{5.45}$$

where $\Phi_p$ is the golden $p$-ratio, $p \in \{0,1,2,3,...\}$.

Taking into consideration (5.45), we will name the resistive divisors in Fig. 5.1 the "golden" resistive divisors [16].

It is clear that the resistive divisor in Fig. 5.1 sets an infinite number of the different "golden" resistive divisors, because every $p$ "generates" a new "golden" resistive divisor. In particular, for the case $p = 0$ the value of the golden 0-ratio $\Phi_0 = 2$ and the "golden" resistive divisor is reduced to the classic "binary" resistive divisor, based on the resistors $R$ and $2R$.

For the case $p = 1$ the resistors $R1$, $R2$, $R3$ in Fig. 5.1 according to (5.45) take the following values:

$$R1 = \Phi^{-1} R = 0.618R;\ R2 = \Phi^2 R = 2.618R;\ R3 = \Phi R = 1.618R. \tag{5.46}$$

For this choice of the values of the resistors $R1$, $R2$, $R3$, given by (5.45), (5.46), the "golden" resistive divisors in Fig. 5.1 have the following unique electric properties. To find these properties, we will use the following fundamental mathematical relations for the *golden p-proportions* $\Phi_p$:

$$\Phi_p = 1 + \Phi_p^{-p}, \tag{5.47}$$

$$\Phi_p^{p+2} = \Phi_p^{p+1} + \Phi_p, \tag{5.48}$$

which takes the following forms for the cases $p = 0$ ($\Phi_{p=0} = 2$) and $p = 1$ $\left(\Phi_{p=1} = \Phi = \frac{1+\sqrt{5}}{2} = 1.618\right)$, respectively:

$$p = 0: 2 = 1 + 1;\ 2^2 = 2 + 2 \tag{5.49}$$

$$p = 1: \Phi = 1 + \Phi^{-1};\ \Phi^3 = \Phi^2 + \Phi. \tag{5.50}$$

By using (5.40), (5.45) and the identity (5.48), we can deduce the value of equivalent resistance $R_{e1}$ of the resistive circuit of the "golden" resistive divisor in Fig. 5.1 to the left and to the right from the "connecting points" 0 and 4. In general case of $p$ ($p \geq 1$) the formula looks as follows:

$$R_{e1} = \frac{R2 \times R3}{R2 + R3} = \frac{\Phi_p^{p+1} R \times \Phi_p R}{\Phi_p^{p+1} R + \Phi_p R} = \frac{\Phi_p^{p+2} R^2}{\left(\Phi_p^{p+1} + \Phi_p\right) R} = R. \tag{5.51}$$

Note that we have simplified the formula (5.51) by using the identity (5.48).

By using (5.41) and (5.47), we can calculate the value of the equivalent resistance of $R_{e2}$ as follows:

$$R_{e2} = \Phi_p^{-p} R + R = \left(\Phi_p^{-p} + 1\right) R = \Phi_p R. \tag{5.52}$$

Thus, according to (5.52) the equivalent resistance of the resistive circuit of the "golden" resistive divisor to the left or to the right of the "connecting points" 0, 1, 2, 3, 4 is equal to $\Phi_p R$, where $\Phi_p$ is the golden $p$-proportion. This fact can be used for the calculation of the equivalent resistance $R_{e3}$ of the "golden" resistive divisor in the "connecting points" 0, 1, 2, 3, 4. In fact, the equivalent resistance $R_{e3}$ can be calculated as the resistance of the resistive circuit, which consists of the parallel connection of the "vertical" resistor $R2 = \Phi_p^{p+1} R$ and the two "lateral" resistors with the resistance $\Phi_p R$. But because according to (5.51) the equivalent resistance of the parallel connection of the resistors $R2 = \Phi_p^{p+1} R$ and $R3 = \Phi_p R$ is equal to $R$, then the equivalent resistance $R_{e3}$ of the "golden" resistive divisor in each "connecting point" can be calculated by the formula:

$$R_{e3} = \frac{\Phi_p R \times R}{\Phi_p R + R} = \frac{\Phi_p R^2}{\left(\Phi_p + 1\right) R} = \frac{1}{1 + \Phi_p^{-1}} R. \tag{5.53}$$

Note that for the case $p = 0$ (the "binary" resistive divisor) $\Phi_{p=0} = 2$ and then the expression (5.53) is reduced to (5.43). For the case $p = 1$ the formula (5.53) is reduced to the following formula:

$$R_{e3} = \frac{1}{1 + \Phi^{-1}} R = \frac{1}{\Phi} R = \Phi^{-1} R. \tag{5.54}$$

Let us calculate now the voltage transmission coefficient between the adjacent "connecting points" of the "golden" resistive divisor. For this purpose we connect the generator of the standard electric current $I$ to one of the "connecting points," for example, to point 2. Then according to Ohm's law, the following electrical voltage appears in this point:

$$U = \frac{1}{1 + \Phi_p^{-1}} RI. \tag{5.55}$$

Note that for the case $p = 0$ $\Phi_{p=0} = 2$ and the formula (5.55) is reduced to the following formula:

$$U = \frac{1}{1 + 2^{-1}} RI = \frac{1}{1 + \frac{1}{2}} RI = \frac{2}{3} RI, \tag{5.56}$$

which coincides with the formula (5.44) for the "binary" resistive divisor.

Let us calculate now the electrical voltage in the adjacent "connecting points" 3 and 1. The voltages in points 3 and 1 can be calculated as a result of linking the voltage $U$, given by (5.55), to the resistive circuit, which consists of the sequential connection of the "horizontal" resistor $R1 = \Phi_p^{-p} R$ and the resistive circuit with the equivalent resistance $R$. Then, for this case the electrical current $I$, which appears in the resistive circuit to the left and to the right of the "connecting point" 2, will be equal to

$$I = \frac{U}{R1 + R} = \frac{U}{\left(\Phi_p^{-p} + 1\right)R} = \frac{U}{\Phi_p R}. \quad (5.57)$$

If we multiply the electrical current (5.57) by the equivalent resistance $R$, we get the following value of the electrical voltage in the adjacent "connecting points" 3 and 1:

$$\frac{U}{\Phi_p}. \quad (5.58)$$

This means that the voltage transmission coefficient between the adjacent "connecting points" of the "golden" resistive divisor in Fig. 5.1 is equal to the reciprocal of the golden $p$-proportion $\Phi_p$!

Thus, the "golden" resistive divisor in Fig. 5.1, based on the golden $p$-proportions is quite real electrical circuits. It is clear that the above theory of the "golden" resistive divisors [16] can become a new source for the development of the "digital metrology" and analog-to-digit and digit-to-analog converters.

## 5.6. Digit-to-analog (DAC) and analog-to-digit converters (ADC) based on the "golden" resistive divisors

### 5.6.1. *The "golden" digit-to-analog converters*

The electrical circuit of the "golden" DAC, based on the "golden" resistive divisor in Fig. 5.1, is shown in Fig. 5.2.

Note that the "golden" DAC in Fig. 5.2 consists of 5 digits. However the number of the DAC digits may be increased to some arbitrary $n$ by extending the "golden" resistive divider to the left and to the right.

The "golden" DAC in Fig.5,2 contains 5 ($n$ in the general case) generators of the standard electrical current $I_0$ and 5 ($n$ in the general case) electrical current keys $K_0 - K_4$. The key states are controlled by the binary digits of the golden $p$-proportion code $a_4a_3a_2a_1a_0$. For the case $a_i = 1$ the key $K_i$ is closed, for the case $a_i = 0$ is open ($i = 0, 1, 2, \ldots, n$).

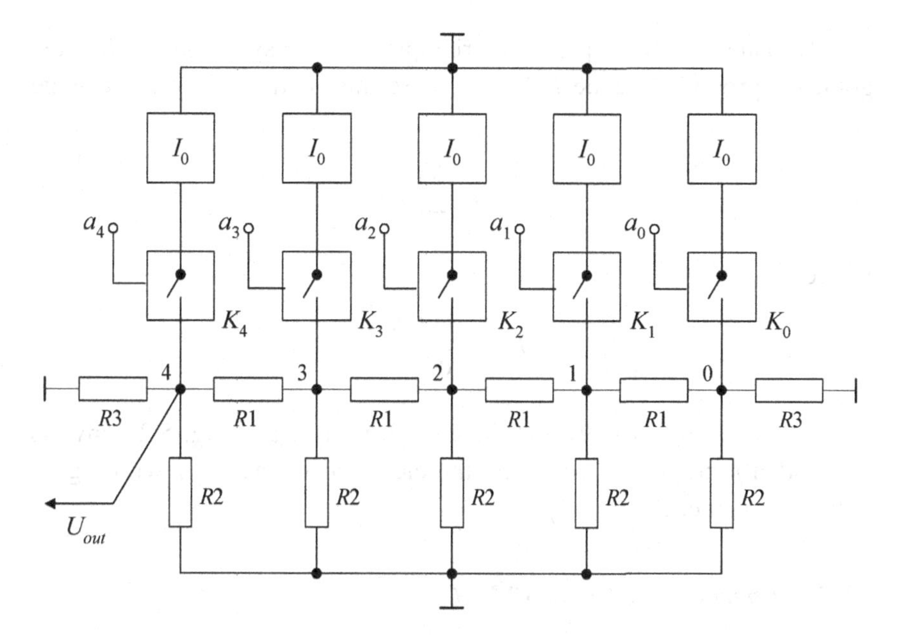


Figure 5.2. The "golden" DAC.

One can show that the closed key $K_i$ results in the following voltage in the $i$th "connecting point" of the resistive divider:

$$U_i = \beta_p I_0 R,$$

where

$$\beta_p = \frac{1}{1+\Phi_p^{-1}}.$$

As the potential $U_i$ is passed from the $i$th point to the $(i + 1)$th point with the transmission coefficient $\frac{1}{\Phi_p}$, the following voltage appears at the DAC output:

$$U_{out} = \frac{\beta_p I_0 R}{\Phi_p^{n-l-1}} = \frac{\beta_p I_0 R}{\Phi_p^{n-1}} \times \Phi_p^i.$$

By using the superposition principle, it is easy to show that the golden $p$-proportion code $a_{n-1}a_{n-2}...a_0$ results in the following voltage $U_{out}$:

$$U_{out} = B_p \sum_{i=0}^{n-1} a_i \Phi_p^i, \tag{5.59}$$

where

$$B_p = \frac{\beta_p I_0 R}{\Phi_p^{n-1}}.$$

It follows from (5.59) that the electrical circuit in Fig. 5.2 converts the golden $p$-proportion code into the electrical voltage $U_{out}$ with regard to the constant coefficient $B_p$.

### 5.6.2. *Checking the "golden" DAC*

In the measurement practice there is a necessity to check the DAC linearity at the production and operation process. For the classic binary DAC the following correlation for checking the DAC linearity is used:

$$2^n = \sum_{i=0}^{n-1} 2^i + 1.$$

The mathematical properties of the golden $p$-proportion codes (5.6) are providing very wide possibilities for checking the "golden" DAC linearity. In particular, the linearity checking of the "golden" DAC, based on the classic golden ratio ($p = 1$), is reduced to the checking of the following relations:

$$\Phi^n = \Phi^{n-1} + \Phi^{n-2} = \Phi^{n-1} + \Phi^{n-3} + \Phi^{n-4} = \Phi^{n-1} + \Phi^{n-3} + \Phi^{n-5} + \Phi^{n-6} \tag{5.60}$$

The checking is fulfilled in the following manner. We have to check that the output voltage of the "golden" DAC in Fig. 5.2 will be constant for the following input "golden" representations, which corresponds to (5.60):

$$\begin{matrix} 1 & 0 & 0 & 0 & 0 & 0 & 0 \\ 0 & 1 & 1 & 0 & 0 & 0 & 0 \\ 0 & 1 & 0 & 1 & 1 & 0 & 0 \\ 0 & 1 & 0 & 1 & 0 & 1 & 1 \end{matrix}.$$

Note that the different input "golden" representations are formed from the top "golden" representation 1000000 by means of the *devolutions*.

### 5.6.3. *Algorithm of the "golden" ADC*

The algorithm and structural scheme of the "golden" ADC do not differ from the algorithm and the structural scheme of the classic binary ADC. However, the golden ratio additive properties (5.24) $\Phi_p^i = \Phi_p^{i-1} + \Phi_p^{i-p-1}$ and (5.26) $\Phi^i = \Phi^{i-1} + \Phi^{i-2}$, which connect the adjacent digit weights of the golden proportion code (5.6), give a number of interesting technical advantages of the "golden" ADC.

Let us convert the analogous magnitude of $X$ in the range:

$$0 \le X < \Phi^n$$

into the $n$-digit golden ratio 1-code by using the digit-by-digit algorithm of the analog-to-digit conversion. Then the digit-by-digit "golden" algorithm may be considered as the process of the sequential representation of value $X$ and all appearing reminders $r_1, r_2, \ldots, r_n$ in the following forms:

*The first step*: $X = \Phi^{n-1} + r_1$, where $0 \le r_1 < \Phi^{n-2}$.

*The second step*: $r_1 = \Phi^{n-3} + r_2$, where $0 \le r_2 < \Phi^{n-4}$.

*The third step*: $r_2 = \Phi^{n-5} + r_3$, where $0 \le r_3 < \Phi^{n-6}$.

It follows from this consideration that the output code of the "golden" ADC is represented in the *minimal form*. It means that the process of the "golden" analog-to-digital conversion is checked in accordance with the *minimal form*.

### 5.6.4. *Self-correcting "golden" ADC*

There is a problem to guarantee the temperature and prolonged in time stability for the high-reliable control systems. Because ADC and DAC are very important devices of high-reliable control systems for many complicated technological objects, therefore designing the self-correcting ADC and DAC is one of the most important areas of applications of the Fibonacci and golden ratio codes [14–16].

While the faults and failures of the digital components of computers and microprocessors (for example, flip-flops and logic gates) are the main cause of non-reliability of the digital systems, the deviations of parameters of the analog elements ADC and DAC from their standard values are the main cause of the informational non-stability of measurement systems. These deviations depend on different interior and exterior factors ("aging" of elements, temperature influences, technological errors etc.) and they are usually the "slow" time functions. In designing the exact measurement systems there is a problem to decrease the requirements to the technological exactness of the analog elements and eliminate such difficult technological procedures as the laser "tuning" of the analog elements. The solution of this problem is realized by applying *principle of self-correction*.

The Fibonacci and golden proportion codes allow one to apply the *principle of self-correction* to improve the exactness and metrological stability of ADC and DAC. At the realization of the "golden" and Fibonacci self-correcting ADC and DAC, the most important advantage is the correction of the non-linearity of transfer function of the "golden" resistive divisor.

As it is known, the "binary" resistive divisor has the following drawback. Let us consider the situation, which can occur at changing two adjacent binary words for ideal and erroneous digit weights (see Table 5.10).

Table 5.10. Two adjacent binary words in the "binary" DAC ($p = 1$).

| $i$ | 9 | 8 | 7 | 6 | 5 | 4 | 3 | 2 | 1 |
|---|---|---|---|---|---|---|---|---|---|
| $W_i$ (*ideal weights*) | 256 | 128 | 64 | 32 | 16 | 8 | 4 | 2 | 1 |
| $W_i$ (*erroneous weigths*) | $\boxed{260}$ | 128 | 64 | 32 | 16 | 8 | 4 | 2 | 1 |
| $A = 2^9 - 1$ (*for ideal and erroneous weigths*) | 0 | 1 | 1 | 1 | 1 | 1 | 1 | 1 | 1 |
| $B = 2^9$ (*for ideal weigths*) | 1 | 0 | 0 | 0 | 0 | 0 | 0 | 0 | 0 |
| $B = 2^9 + 4$ (*for erroneous weigths*) | 1 | 0 | 0 | 0 | 0 | 0 | 0 | 0 | 0 |

For erroneous case, when the weight of the 9th digit $2^9 = 256$ deviates from the standard value on 4 units and is equal to 260, there arises a "break" in the transfer function, if the binary word $A = 011111111$ at the DAC input is replaced on the binary word $B = 1000000000$.

The violation of the *linearity* of the transfer function of the "binary" resistive divisor leads to a fatal uncorrected error of the "binary" DAC.

Let us consider the Fibonacci DAC, in which the digit weights are Fibonacci numbers: 1, 1, 2, 3, 5, 8, 13, 21, 34 (see Table 5.11).

Table 5.11. Correction of erroneous weight in Fibonacci DAC.

| $i$ | 9 | 8 | 7 | 6 | 5 | 4 | 3 | 2 | 1 |
|---|---|---|---|---|---|---|---|---|---|
| $F_i$ (*erroneous weights*) | $\boxed{37}$ | 21 | 13 | 8 | 5 | 3 | 2 | 1 | 1 |
| $A = F_9 - 1 = 33$ (*for ideal and erroneous weigths*) | 0 | 1 | 0 | 1 | 0 | 1 | 0 | 1 | 0 |
| $B = F_9 = 34$ (*for ideal weigths*) | 0 | 1 | 0 | 1 | 0 | 1 | 0 | 1 | 0 |
| $B = F_9 + 3 = 34 + 3$ (*for erroneous weigths*) | 1 | 0 | 0 | 0 | 0 | 0 | 0 | 0 | 0 |
| $B_{cor} = F_9 = 34$ (*for erroneous weigths*) | $\boxed{\times}$ | 1 | 0 | 1 | 0 | 1 | 0 | 1 | 1 |

We consider the case, when the input Fibonacci representations are *minimal forms* (two bits of 1 do not occur together). The 9th-digit Fibonacci representations $A = 010101010 = 33$ and $B = 100000000 = 34$ are the adjacent Fibonacci representations (for the case of the ideal weights). However, for the case of the erroneous digit weight $\boxed{37}$ (instead 34) we can't use the erroneous digit weight $\boxed{37}$, if we chose the Fibonacci representations

$$34 = 010101011 = 010101100 = 010110000 = 011000000,$$

which is not *minmal form*, as the adjacent to the *minmal form* $A$ = 010101010 = 33. The next adjacent Fibonacci representations are the following:

$$35 = 011000001 = 011000010;$$

$$36 = 011000011 = 011000100.$$

After this, we can use the erroneous digit weight $\boxed{37}$ and represent the next input adjacent Fibonacci code words as follows:

$$37 = 011000101 = 011000110 = 011001000.$$

The main conclusion from the above examples consists in the fact that by using Fibonacci code, we can realize exact digital-to-analog conversion on the resistive divisor with wrong weights. This is achieved thanks to the *multiplicity* of number representation in the Fibonacci or "golden" code.

The other way is to determine the erroneous weight $\boxed{37}$. It turns out that the easiest way to solve this task is to use the Fibonacci or "golden" ADC.

Let us demonstrate how we can determine the erroneous digit weight $\boxed{37}$ in Fig. 4.10 on the example of the 9-digit Fibonacci ADC. The latter uses the Fibonacci numbers 1, 1, 2, 3, 5, 8, 13, 21, 34 as the ideal digit weights. Let us suppose that the Fibonacci weight of the higher digit has the deviation from its standard value in 3 units. Thus the "real" weights of the Fibonacci ADC are equal to: 1, 1, 2, 3, 5, 8, 13, 21, $\boxed{37}$. Let us send some analog value 39, which is more than the erroneous weight $\boxed{37}$ and lesser than the next Fibonacci weight 55, to the ADC-input. Let us fulfill now the analog-to-digit conversion of the input value 39 twice: firstly with the use of the higher erroneous weight $\boxed{37}$ and secondly without it. As a result, we get the two code combinations (Table 5.12).

Table 5.12. Measuring the erroneous weight 37.

| $i$ | $\to$ | 9 | 8 | 7 | 6 | 5 | 4 | 3 | 2 | 1 |
|---|---|---|---|---|---|---|---|---|---|---|
| $W_i\left(with\ error\right)$ | $\to$ | $\boxed{37=34+3}$ | 21 | 13 | 8 | 5 | 3 | 2 | 1 | 1 |
| $A$ | = | 1 | 0 | 0 | 0 | 0 | 0 | 1 | 0 | 0 |
| $B$ | = | 0 | 1 | 1 | 0 | 0 | 1 | 0 | 1 | 0 |
| $D$ | = | 0 | 0 | 0 | 0 | 0 | 1 | 0 | 0 | 0 |

If we interpret the code combinations

$$A = 100000100 \text{ and } B = 011001010$$

as the Fibonacci representations with the "ideal" weights 1, 1, 2, 3, 5, 8, 13, 21, 34, we can find that their difference

$$D = 00001000 = 3.$$

Hence, we have obtained the value of the deviation $D = 3$ of the higher erroneous digit weight $\boxed{37}$ from the standard value 34.

In general, all the digit weights of the Fibonacci resistive divisor have some deviations from ideal weights (Fibonacci numbers or the golden proportion powers). But for the same relative error of manufacturing digit weights, the greatest absolute deviations from the ideal values (Fibonacci numbers or the golden proportion powers) have the weights of the highest digits. This experimental fact gives us the right to divide all the digit weights into two groups: the group of the lower digits, which are considered as "conditionally ideal," and the group of the highest digits, in which the deviation from the ideal value exceeds the predetermined error of analog-to-digital conversion (see Table 5.13).

Table 5.13. The Fibonacci weights with ideal and erroneous weights.

| $i$ | 9 | 8 | 7 | 6 | 5 | 4 | 3 | 2 | 1 |
|---|---|---|---|---|---|---|---|---|---|
| $W_{ideal} = F_i$ (*the ideal weights*) | 34 | 21 | 13 | 8 | 5 | 3 | 2 | 1 | 1 |
| $W_{non-ideal}$ (*the weights with errors*) | $34 \pm \Delta_9$ | $21 \pm \Delta_8$ | $13 \pm \Delta_7$ | $8 \pm \Delta_6$ | 5 | 3 | 2 | 1 | 1 |

Here the digits of the numbers 1–5 are the "conditionally ideal" weights (the Fibonacci numbers 1, 1, 2, 3, 5), but the digits of the number 6–9 have the weights with errors: $8 \pm \Delta_6$, $13 \pm \Delta_7$, $21 \pm \Delta_8$, $34 \pm \Delta_9$.

The procedure for determining the deviation of the weights of the highest digits from their ideal values (Fibonacci numbers) starts from the 6th digit and ends with the 9th digit. For this purpose, we send consistently to the input of ADC, some analog values ($X_i$ ($i = 6, 7, 8, 9$)), satisfying the following conditions:

$$W_i < X_i < W_{i+1},$$

where $W_i$, $W_{i+1}$ are the adjacent erroneous digit weights.

Further, at each step of the above procedure, we calculate the weight deviation of the appropriate digit from its ideal value (the Fibonacci numbers). The values of the deviations are stored in the memory of the microprocessor of the Fibonacci or "golden" ADC.

After completing the procedure for determining the deviations of the digit weights from their ideal values (Fibonacci numbers), the main mode of analog-to-digital conversion has been switched on in the Fibonacci or "golden" ADC. After completing the process of analog-to-digital conversion, the microprocessor calculates the true value of the conversion result, taking into consideration the deviations of digit weights from their ideal values (Fibonacci numbers or the powers of the golden proportion), and then converts the result of analog-to-digital conversion into the binary code.

This procedure is used in the manufacturing process for automatic adjusting of ADC to the given exactness and then it is periodically repeated depending on the aging of the elements and external factors, for example, temperature changes which lead to the deviation of the digit weights from their ideal values (Fibonacci numbers or the powers of the golden proportion).

In the Special Design Bureau "Module" of the Vinnytsia Technical University (Ukraine), under author's scientific leadership, several modifications of self-correcting ADCs and DACs were developed, in which the above procedure for correcting the deviations of the digit weights from their ideal values (Fibonacci numbers or the golden ratio powers) was realized.

The self-correcting 17-digit ADC, based on the Fibonacci numbers, was one of the best engineering developments, which was designed and produced in the Special Design Bureau "Module" [92, 93] (Fig. 5.3).

Figure 5.3. 17-digit self-correcting ADC.

ADC in Fig. 5.3 had the following technical parameters:

1. The number of digits — 17 (16 digital and one sign digit)
2. Conversion time — 15 microseconds
3. Total error — 0,006%
4. Linearity error — 0,003%
5. Frequency range — 25 kHz
6. Operating temperature range — 20± 30°C.

The correction system, built in the ADC, allows to correct the *zero drift*, the *linearity* of the AD-conversion, which is fulfilled by traditional methods, and most importantly, to correct the deviations of the digit weights from their nominal values (Fibonacci numbers or powers of the golden proportion).

According to the opinion of the Soviet well-known metrological firms, the Soviet electronic industry had not produced ADC with such high technical parameters.

## 5.7. Conclusion

1. The prominent American scientist, physicist and mathematician John von Neumann (1903–1957), together with his colleagues from the Prinstone Institute for Advanced Study Goldstein and Berks after careful analysis of the strengths and weaknesses of the first electronic computer ENIAC gave definite preference to the binary system as a universal way of coding of data in electronic computers. “Von Neumann principles” were the beginning of a modern computer revolution based on the binary system. However, the classical binary code has zero code redundancy what excludes a possibility detecting any errors in computer structures. Russian scientist Yaroslav Khetagurov called this danger “Trojan horse” of binary system. Because of the "Trojan Horse" phenomenon, humanity becomes a hostage to the binary system for the case of mission-critical applications. *From here, it follows the conclusion that the binary system is unacceptable for designing computing and measuring systems for mission-critical applications.*

2. The theory of numeral systems with irrational bases, described in this book, is a new direction in the field of coding theory, intended for increasing informational reliability and noise immunity of specialized computing and measuring systems for mission-critical applications. In their origin, these positional numeral systems date back to the Babylonian positional numeral system with base 60, which became the precursor of all known positional numeral systems, in particular, the decimal, which is the basis of the initial mathematical education, and the binary, which is the basis of modern computers. These numeral systems, based on the positional principle, are a generalization of classical binary and ternary numeral systems and preserve all advantages of the binary and ternary systems. However, code redundancy in combination

to the positional principle is their main advantage in comparison with the classical binary and ternary numeral systems. This direction does not set itself the task of replacing the classical binary system in those cases where the use of binary (or ternary) systems does not threaten an appearance of technological disaster and where informational reliability and noise immunity can be ensured by traditional means. The main task of numeral systems with irrational bases is to prevent or significantly reduce the probability of "false signals" at the output of informational systems which can lead to social or technological disasters. These numeral systems return mathematics to the period of its origin (Babylon, Ancient Egypt) and unite the Babylonian positional principle of representing numbers and Egyptian "doubling method" with the "golden section," the main mathematical achievement of the ancient Greeks in "mathematical harmony." They can lead to a revision of the existing number theory, which goes back to Euclid's *Elements* and lead to the creation of the "golden" theory of natural numbers.

# Bibliography

[1] Kolmogorov, A.N. *Mathematics in its Historical Development.* Moscow: Science, (1991) (Russian).

[2] Bashmakova, J.G., Youshkevich, A.P. An origin of the numeral systems. Encyclopedia of Elementary Arithmetics. Book 1. Arithmetic. Moscow-Leningrad: Gostekhizdat (1951) (Russian).

[3] Losev A. *The history of philosophy as a school of thought.* Journal "Communist," (1981), №. 11 (Russian).

[4] Stakhov A.P. *The Mathematics of Harmony. From Euclid to Contemporary Mathematics and Computer Science.* Assisted by Scott Olsen. World Scientific, (2009).

[5] Pythagoreanism. From Wikipedia, the free encyclopedia https://en.wikipedia.org/wiki/Pythagoreanism

[6] Pospelov, D.A. *Arithmetic Foundations of Computers.* Moscow: High School, (1970) (Russian).

[7] *Mission critical.* From Wikipedia, the free encyclopedia https://en.wikipedia.org/wiki/Mission_critical

[8] Khetagurov, J.A. *Ensuring the national security of real-time systems, BC / NW* (2009), №. 2 (15).

[9] *Error detection and correction.* From Wikipedia, the free encyclopedia https://en.wikipedia.org/wiki/Error_detection_and_correction

[10] Bergman, G. *A number system with an irrational base.* Mathematics Magazine, 1957, No. 31.

[11] Stakhov, A.P. *Synthesis of optimal algorithms for analog-to-digital conversion.* Doctoral thesis, Kiev Institute of Civil Aviation Engineers (1972) (Russian).

[12] Stakhov, A.P. *Redundant binary positional numeral systems.* In "Homogenous digital computer and integrated structures." Taganrog Radio University, No 2 (1974) (Russian).

[13] Stakhov, A.P. *A use of natural redundancy of the Fibonacci number systems for computer systems control.* Automation and Computer Systems (1975) No. 6 (Russian).

[14] Stakhov, AP. *Introduction into Algorithmic Measurement Theory*. Moscow: Soviet Radio (1977) (Russian).

[15] Stakhov, A.P. *Codes of the Golden Proportion*. Moscow: Radio and Communication (1984) (Russian).

[16] Stakhov, A.P. *Digital Metrology on the basis of the Fibonacci codes and Golden Proportion Codes*. In "Contemporary Problems of Metrology." Moscow: Moscow Machine-building Institute (1978) (Russian).

[17] Stakhov, A.P. *Fibonacci and "Golden" Ratio Codes*. In "Fault-tolerant Systems and Diagnostic FTSD-78," Gdansk (1978).

[18] Stakhov, A.P. *The golden mean in the digital technology*. Automation and Computer Systems (1980) No. 1 (Russian).

[19] Stakhov, A.P. *Algorithmic measurement theory and fundamentals of computer arithmetic*. Measurement. Control. Automation (1988) No. 2 (Russian).

[20] Stakhov, A.P. *The Golden Proportion Principle: Perspective Way of Computer Progress*. Bulletin of the National Academy of Sciences of Ukraine (1990) No. 1-2 (Ukrainian).

[21] Stakhov, A.P. *The Golden Section and Science of System Harmony*. Bulletin of the National Academy of Sciences of Ukraine (1991) No. 12 (Ukrainian).

[22] Stakhov, A.P. *Algorithmic Measurement Theory: A General Approach to Number Systems and Computer Arithmetic*. Control Systems and Computers (1994) No. 4-5 (Russian).

[23] Luzhetsky, V.A., Stakhov, A.P., Wachowski, V.G. *Noise-immune Fibonacci computers*. The brochure "Noise-immune codes. Fibonacci Computer". Moscow: Knowledge (1989) (Russian).

[24] Stakhov, A.P. *The Golden Section in the Measurement Theory*. Computers & Mathematics with Applications (1989) Vol. 17, No. 4-6.

[25] Stakhov, A.P. *The Golden Section and Modern Harmony Mathematics*. Applications of Fibonacci Numbers, Kluwer Academic Publishers (1989) Vol. 7.

[26] Stakhov, A.P. *A generalization of the Fibonacci Q-matrix*. Reports of the National Academy of Sciences of Ukraine (1999) No. 9.

[27] Stakhov, A.P. *Generalized golden sections and a new approach to the geometric definition of the number*. Ukrainian Mathematical Journal (2004) Vol. 56, No. 8 (Russian).

[28] Stakhov, AP. *Brusentsov's ternary principle, Bergman's number system and ternary mirror-symmetrical arithmetic*. The Computer Journal (2002) Vol. 45, No. 2: 222-236.

[29] Stakhov, A.P. *Fibonacci matrices, a generalization of the "Cassini formula," and a new coding theory*. Chaos, Solitons & Fractals (2006) Vol. 30, Issue 1.

[30] Stakhov, AP. *The generalized golden proportions, a new theory of real numbers, and ternary mirror-symmetrical arithmetic*. Chaos, Solitons & Fractals (2007) Vol. 33, Issue 2.

[31] Stakhov, A.P. *The Mathematics of Harmony: Clarifying the Origins and Development of Mathematics*. Congressus Numerantium (2008) Vol. 193.

[32] Stakhov, A.P. *A generalization of the Cassini formula*. Visual Mathematics, (2012), No. 2. http://www.mi.sanu.ac.rs/vismath/stakhovsept2012/cassini.pdf

[33] Stakhov, A.P. *The "Golden" Number Theory and New Properties of Natural Numbers*. British Journal of Mathematics & Computer Science (2015) Vol. 11, No. 6.

[34] Stakhov A.P. *Proclus Hypothesis*. British Journal of Mathematics & Computer Science (2016) Vol. 13, No. 6.

[35] Stakhov, A.P. *Fibonacci p-codes and Codes of the Golden p-proportions: New Informational and Arithmetical Foundations of Computer Science and Digital Metrology for Mission-Critical Applications*. British Journal of Mathematics & Computer Science (2016) Vol. 17, No. 1.

[36] Stakhov, A.P. *The importance of the Golden Number for Mathematics and Computer Science: Exploration of the Bergman's system and the Stakhov's Ternary Mirror-symmetrical System (Numeral Systems with Irrational Bases)*. British Journal of Mathematics & Computer Science (2016) Vol. 18, No. 3.

[37] *Reduction method of p-Fibonacci code to the minimal form and device for its realization*. Patent certificate of USA No. 4187500.

[38] *Device for reduction of p-Fibonacci codes to the minimal form*. Patent certificate of USA No. 4290051.

[39] *Reduction method of p-Fibonacci code to the minimal form and device for its realization*. Patent certificate of England No. 1543302.

[40] *Device for reduction of p-Fibonacci codes to the minimal form*. Patent certificate of England No. 2050011.

[41] *Reduction method of p-Fibonacci code to the minimal form and device for its realization*. Patent certificate of Germany No. 2732008.

[42] *Device for reduction of p-Fibonacci codes to the minimal form*. Patent certificate of Germany No. 2921053.

[43] *Reduction method of p-Fibonacci code to the minimal form and device for its realization*. Patent certificate of Japan No. 1118407.

[44] *Reduction method of p-Fibonacci code to the minimal form and device for its realization*. Patent certificates of France No. 7722036, No. 2359460.

[45] *Device for reduction of p-Fibonacci codes to the minimal form*. Patent certificates of France No. 7917216, No. 2460367.

[46] *Reduction method of p-Fibonacci code to the minimal form and device for its realization*. Patent certificate of Canada No. 1134510.

[47] *Device for reduction of p-Fibonacci codes to the minimal form*. Patent certificate of Canada No 1132263.

[48] *Reduction method of p-Fibonacci code to the minimal form and device for its realization*. Patent certificate of Poland No. 108086.

[49] *Reduction method of p-Fibonacci code to the minimal form and device for its realization.* Patent certificate of DDR No. 150514.

[50] Klein, Felix *Lectures on the Icosahedron; and the Solution of Equations of the Fifth Degree*, 2nd revised edition, New York (2014).

[51] Grimm, G.D. *Proportionality in Architecture*. Leningrad-Moscow: Publishing House "ONTI" (1935) (Russian).

[52] Soroko, E.M. *Structural harmony of systems*. Minsk: Science and Technology (1984) (Russian).

[53] Vorobyov, N.N. *Fibonacci Numbers*. Moscow: Nauka, (1961) (Russian).

[54] Hoggatt, V.E. *Fibonacci and Lucas Numbers*. Palo Alto, CA: Houghton-Mifflin, (1969).

[55] Koshy, T. *Fibonacci and Lucas Numbers with Applications*, 2nd edition. John Wiley & Sons, Inc. (2017).

[56] Kann, Charles H. *Pythagoras and Pythagoreans. A Brief History*. - Hackett Publishing Co, Inc. (2001).

[57] Zhmud, L. *The origin of the History of Science in Classical Antiquity*. Walter de Gruyter (2006).

[58] Smorinsky, Craig. *History of Mathematics. A Supplement*. Springer, (2008).

[59] Bodnar, O. Y. *The Golden Section and Non-Euclidean Geometry in Nature and Art*. Lviv: Svit (1994) (Russian).

[60] *Pell number*. From Wikipedia, the free encyclopedia https://en.wikipedia.org/wiki/Pell_number

[61] Polya, G. *Mathematical Discovery. On understanding, learning and teaching problem solving*. New York – London: Volume I (1962); Volume II (1965).

[62] *Turing, Alan*. From Wikipedia, the free encyclopedia https://en.wikipedia.org/wiki/Alan_Turing

[63] Swinton, *J. Fibonacci phyllotaxis: Turing's problem* (2002) www.swintons.net/jonathan/Turing/fibonacci.htm

[64] Turing, A.M. *The Chemical Basis of Morphogenesis*. Philosophical Transactions of the Royal Society of London (1952) Vol. B 237.

[65] Turing, A.M. *The morphogenetic theory of phyllotaxis*. In Saunders (1992).

[66] Depman, J.J. *A History of Arithmetic*. Moscow: Uchpedgiz (1959) (Russian).

[67] Neugebauer O. *Vorlesungen uber Geshichte der Antices Mathematishes Wissenshaftes*. Ester Band. Vorgrieschishen Matematik. Berlin: Verlag von Jullius Spinger (1931).

[68] *Egyptian calendar*. From Wikipedia, the free encyclopedia https://en.wikipedia.org/wiki/Egyptian_calendar

[69] *Binary number*. From Wikipedia, the free encyclopaedia https://en.wikipedia.org/wiki/Binary_number

[70] Burks, A.W., Goldstine, H.H., von Neumann, John. *Preliminary discussion of the logical design of an electronic computing instrument* http://www.cs.unc.edu/~adyilie/comp265/vonNeumann.html

[71] Brousentsov, N.P. *Computer Machine "SETUN" of Moscow State University*. In "New Development of Computer Technology." Kiev: Kiev Cybernetics Institute of the Ukrainian Academy of Sciences (1960) (Russian).

[72] Knuth D. E. *The Art of Computer Programming. Volume 1. Fundamental Algorithms (Third edition).* Massachusetts: Addison-Wesley (1997).

[73] *Digital Signal Processors Based on Ternary Code* http://www.ci.ru/inform13_08/p_23.htm

[74] *Physicists have created a nano-memory by using ternary logic* http://www.moyaufa.ru/35590/1/view/news.html

[75] *George Bergman.* From Wikipedia, the free encyclopaedia https://en.wikipedia.org/wiki/George_Bergman

[76] *Golden ratio base.* From Wikipedia, the free encyclopaedia https://en.wikipedia.org/wiki/Golden_ratio_base

[77] *Phi Number System.* From WolframMathWorld http://mathworld.wolfram.com/PhiNumberSystem.html

[78] Shanin, N.A. *Recursive Mathematical Analysis and Calculus of Arithmetic Equations by R.L. Goodstein.* Introduction article to the book "Recursive Mathematical Analysis" by R.L. Goodstein. Moscow: Science (1970) (Russian).

[79] Markov, A.A. *The Logic of Constructive Mathematics.* Moscow: Knowledge (1972).

[80] *Zeckendorf., Edouard.* From Wikipedia, the free encyclopaedia https://en.wikipedia.org/wiki/Edouard_Zeckendorf

[81] *Genetic code.* From Wikipedia, the free encyclopaedia https://en.wikipedia.org/wiki/Genetic_code

[82] Kautz, W.H. *Fibonacci codes for synchronization control.* IEEE Transactions on Information Theory (1965) Vol. 11, Issue 2.

[83] Borisenko A.A., Stakhov A.P. *About the Counting Method for the Fibonacci Code.* Journal of the Sumy State University (2011) No.3 (Russian).

[84] Wishnjakov, Y.M. *Development of Principles of Design and Research of Counting Devices in the Fibonacci p-codes.* PhD thesis. Taganrog Radio Engineering Institute (1977) (Russian).

[85] Licomendes, P. and Newcomb, R. *Multilevel Fibonacci Conversion and Addition,* The Fibonacci Quarterly (1984) Vol. 22, No. 3.

[86] Ligomenides, P. and Newcomb, R. *Equivalence of some Binary, Ternary, and Quaternary Fibonacci Computers.* Proceeding of the Eleventh International Symposium on Multiple-Valued Logic, Norman, Oklahoma, (1981).

[87] Ligomenides, P. and Newcomb, R. *Complement Representations in the Fibonacci Computer,* Proceedings of the Fifth Symposium ob Computer Arithmetic, Ann Arbor, Michigan (1981).

[88] Newcomb, R. *Fibonacci Numbers as a Computer Base.* Conference Proceedings of the Second Inter-American Conference on Systems and Informatics, Mexico City, (1974).

[89] Hoang, V.D. *A Class of Arithmetic Burst-Error-Correcting Codes for the Fibonacci Computer.* PhD thesis, University Maryland, (1979).

[90] *Metrology.* From Wikipedia, the free encyclopedia (Russian) https://ru.wikipedia.org/wiki/%D0%9C%D0%B5%D1%82%D1%80%D0%BE%D0%BB%D0%BE%D0%B3%D0%B8%D1%8F#.D0.A6.D0.B5.D0.BB.D0.B8_.D0.B8_.D0.B7.D0.B0.D0.B4.D0.B0.D1.87.D0.B8_.D0.BC.D0.B5.D1.82.D1.80.D0.BE.D0.BB.D0.BE.D0.B3.D0.B8.D0.B8

[91] *Metrology.* From Wikipedia, the free encyclopedia https://en.wikipedia.org/wiki/Metrology

[92] Stakhov, A.P., Azarov, A.D. Moiseev. V.I., Martsenyuk, V.P., Stejskal, V.Y. *The 17-bit Self-correcting ADC.* Devices and Control Systems (1986) №. 1.

[93] Stakhov, A.P., Azarov, A.D. Moiseev. V.I., Stejskal, V.Y. *Analog-to-digital Converters on the Basis of Redundant Numeral Systems.* In "Noise-tolerant codes. Fibonacci computer." Moscow: Science (1986) №. 9 (Russian).

# Index

## SERIES ON KNOTS AND EVERYTHING

ISSN: 0219-9769



---

The Series on Knots and Everything: is a book series polarized around the theory of knots. Volume 1 in the series is Louis H Kauffman's Knots and Physics.

One purpose of this series is to continue the exploration of many of the themes indicated in Volume 1. These themes reach out beyond knot theory into physics, mathematics, logic, linguistics, philosophy, biology and practical experience. All of these outreaches have relations with knot theory when knot theory is regarded as a pivot or meeting place for apparently separate ideas. Knots act as such a pivotal place. We do not fully understand why this is so. The series represents stages in the exploration of this nexus.

Details of the titles in this series to date give a picture of the enterprise.

*Published:*

Vol. 61: *Numeral Systems with Irrational Bases for Mission-Critical Applications*
by A. Stakhov

Vol. 60: *New Horizons for Second-Order Cybernetics*
edited by A. Riegler, K. H. Müller & S. A. Umpleby

Vol. 59: *Geometry, Language and Strategy: The Dynamics of Decision Processes — Vol. 2*
by G. H. Thomas

Vol. 58: *Representing 3-Manifolds by Filling Dehn Surfaces*
by R. Vigara & A. Lozano-Rojo

Vol. 57: *ADEX Theory: How the ADE Coxeter Graphs Unify Mathematics and Physics*
by S.-P. Sirag

Vol. 56: *New Ideas in Low Dimensional Topology*
edited by L. H. Kauffman & V. O. Manturov

Vol. 55: *Knots, Braids and Möbius Strips*
by J. Avrin

Vol. 54: *Scientific Essays in Honor of H Pierre Noyes on the Occasion of His 90th Birthday*
edited by J. C. Amson & L. H. Kauffman

Vol. 53: *Knots and Physics (Fourth Edition)*
by L. H. Kauffman

Vol. 52: *Algebraic Invariants of Links (Second Edition)*
by J. Hillman

More information on this series can also be found at http://www.worldscientific.com/series/skae

Vol. 51: *Virtual Knots: The State of the Art*
by V. O. Manturov & D. P. Ilyutko

Vol. 50: *Topological Library*
*Part 3: Spectral Sequences in Topology*
edited by S. P. Novikov & I. A. Taimanov

Vol. 49: *Hopf Algebras*
by D. E. Radford

Vol. 48: *An Excursion in Diagrammatic Algebra: Turning a Sphere from Red to Blue*
by J. S. Carter

Vol. 47: *Introduction to the Anisotropic Geometrodynamics*
by S. Siparov

Vol. 46: *Introductory Lectures on Knot Theory*
edited by L. H. Kauffman, S. Lambropoulou, S. Jablan & J. H. Przytycki

Vol. 45: *Orbiting the Moons of Pluto*
*Complex Solutions to the Einstein, Maxwell, Schrödinger and Dirac Equations*
by E. A. Rauscher & R. L. Amoroso

Vol. 44: *Topological Library*
*Part 2: Characteristic Classes and Smooth Structures on Manifolds*
edited by S. P. Novikov & I. A. Taimanov

Vol. 43: *The Holographic Anthropic Multiverse*
by R. L. Amoroso & E. A. Ranscher

Vol. 42: *The Origin of Discrete Particles*
by T. Bastin & C. Kilmister

Vol. 41: *Zero to Infinity: The Fountations of Physics*
by P. Rowlands

Vol. 40: *Intelligence of Low Dimensional Topology 2006*
edited by J. Scott Carter et al.

Vol. 39: *Topological Library*
*Part 1: Cobordisms and Their Applications*
edited by S. P. Novikov & I. A. Taimanov

Vol. 38: *Current Developments in Mathematical Biology*
edited by K. Mahdavi, R. Culshaw & J. Boucher

Vol. 37: *Geometry, Language, and Strategy*
by G. H. Thomas

Vol. 36: *Physical and Numerical Models in Knot Theory*
edited by J. A. Calvo et al.

Vol. 35: *BIOS — A Study of Creation*
by H. Sabelli

Vol. 34: *Woods Hole Mathematics — Perspectives in Mathematics and Physics*
*edited by N. Tongring & R. C. Penner*

Vol. 33: *Energy of Knots and Conformal Geometry*
by J. O'Hara

Vol. 32: *Algebraic Invariants of Links*
by J. A. Hillman

Vol. 31: *Mindsteps to the Cosmos*
by G. S. Hawkins

Vol. 30: *Symmetry, Ornament and Modularity*
by S. V. Jablan

Vol. 29: *Quantum Invariants — A Study of Knots, 3-Manifolds, and Their Sets*
by T. Ohtsuki

Vol. 28: *Beyond Measure: A Guided Tour Through Nature, Myth, and Number*
by J. Kappraff

Vol. 27: *Bit-String Physics: A Finite and Discrete Approach to Natural Philosophy*
by H. Pierre Noyes; edited by J. C. van den Berg

Vol. 26: *Functorial Knot Theory — Categories of Tangles, Coherence, Categorical Deformations, and Topological Invariants*
by David N. Yetter

Vol. 25: *Connections — The Geometric Bridge between Art and Science (Second Edition)*
by J. Kappraff

Vol. 24: *Knots in HELLAS '98 — Proceedings of the International Conference on Knot Theory and Its Ramifications*
edited by C. McA Gordon, V. F. R. Jones, L. Kauffman, S. Lambropoulou & J. H. Przytycki

Vol. 23: *Diamond: A Paradox Logic (Second Edition)*
by N. S. Hellerstein

Vol. 22: *The Mathematics of Harmony — From Euclid to Contemporary Mathematics and Computer Science*
by A. Stakhov (assisted by S. Olsen)

Vol. 21: *LINKNOT: Knot Theory by Computer*
by S. Jablan & R. Sazdanovic

Vol. 20: *The Mystery of Knots — Computer Programming for Knot Tabulation*
by C. N. Aneziris

Vol. 19: *Ideal Knots*
by A. Stasiak, V. Katritch & L. H. Kauffman

Vol. 18: *The Self-Evolving Cosmos: A Phenomenological Approach to Nature's Unity-in-Diversity*
by S. M. Rosen

Vol. 17: *Hypercomplex Iterations — Distance Estimation and Higher Dimensional Fractals*
by Y. Dang, L. H. Kauffman & D. Sandin

Vol. 16: *Delta — A Paradox Logic*
by N. S. Hellerstein

Vol. 15: *Lectures at KNOTS '96*
by S. Suzuki

Vol. 14: *Diamond — A Paradox Logic*
by N. S. Hellerstein

Vol. 13: *Entropic Spacetime Theory*
by J. Armel

Vol. 12: *Relativistic Reality: A Modern View*
edited by J. D. Edmonds, Jr.

Vol. 11: *History and Science of Knots*
edited by J. C. Turner & P. van de Griend

Vol. 10: *Nonstandard Logics and Nonstandard Metrics in Physics*
by W. M. Honig

Vol. 9: *Combinatorial Physics*
by T. Bastin & C. W. Kilmister

Vol. 8: *Symmetric Bends: How to Join Two Lengths of Cord*
by R. E. Miles

Vol. 7: *Random Knotting and Linking*
edited by K. C. Millett & D. W. Sumners

Vol. 6: *Knots and Applications*
edited by L. H. Kauffman

Vol. 5: *Gems, Computers and Attractors for 3-Manifolds*
by S. Lins

Vol. 4: *Gauge Fields, Knots and Gravity*
by J. Baez & J. P. Muniain

Vol. 3: *Quantum Topology*
edited by L. H. Kauffman & R. A. Baadhio

Vol. 2: *How Surfaces Intersect in Space — An Introduction to Topology (Second Edition)*
by J. S. Carter

Vol. 1: *Knots and Physics (Third Edition)*
by L. H. Kauffman

Printed in the United States
By Bookmasters